A Guide for the Study of Animals

By Worrallo Whitney and Frederic Lucas and Harold Shinn and Mabel Smallwood

I0484406

PREFACE

The following guide to the study of animals is intended for pupils in secondary schools. It was prepared by the authors at the request of the Biology Round Table, an association composed of the teachers of Biology in the Chicago High Schools, to whom the authors wish to take this opportunity of expressing their appreciation of the many helpful suggestions and criticisms of the manuscript.

The time has passed when a high school course in zoology consists simply of a somewhat simplified edition of a similar course in college. All teachers now recognize that the motivization of any course should be its adaptability to the needs of the student, and that zoology must be taught from the standpoint of the student rather than that of the subject. In preparing this guide, the authors have tried to keep these points in mind.

The matter of presentation, the order of topics, and the choice of material has been much discussed, but the trend of opinion has finally set in toward an ecological rather than a type study of animals; that there should be in the case of young students a brief study of rather a large number of animals to bring out some general biological law, rather than an exhaustive study of a very few types. It is further recognized that the use of a reference library is absolutely essential in connection with and to supplement the laboratory work, as there are some topics beyond the ability of the young student for original investigation as well as impossible in the amount of time usually allotted to the subject in our crowded curricula. Of great importance is the economic side of zoology, especially its bearing upon the applied sciences of medicine, sanitation, household science, and agriculture, and this phase has received special attention in this guide.

The desirability of field work has always been recognized, but the special conditions under which schools must work are so variable as to make any set directions for field work of little value, and so they have in most cases been omitted in this work. Each teacher can easily give such special direction for collecting material and study in the field as the locality of the school and the time available for it shall determine.

Since zoology will probably be the pupil's first laboratory science, the authors have preceded the more formal portion of the manual with a series of short exercises on familiar and easily obtained animals in order to introduce the pupil to the laboratory method and to stimulate his interest, training him at the outset to be constantly on the outlook for specimens and to show him how much may be learned from common things right around him, if he will only use his eyes. We have also begun the more formal portion of the guide with insects, since in the fall they are easily collected and may be studied alive. They illustrate, moreover, the principles of classification and method of using keys and other means of finding out the names of animals. This would seem to be pedagogically sound, for some recent experiments with pupils show that the first question that comes into a child's mind upon seeing a new or strange specimen is "What is it?"

A larger portion of the guide is given to the chordates than is usually the case. The authors also believe that this is correct and in accord with the natural interest of the pupil. It will serve to connect his zoology more closely with his daily experiences.

There is material enough provided to allow the teacher a chance to select that best adapted to his purposes or conditions as well as to provide for those schools that give more than one year to zoology.

CONTENTS

PAGE

CHAPTER I

INTRODUCTORY STUDIES OF LIVING ANIMALS 1

Fly, 1; Maggot, 3; Cockroach, 4; Spider, 5; Cricket, 6; Grasshopper, 7; Butterfly or Moth, 9; Caterpillar, 10; Tussock Moth, 11; Beetle, 13; Damsel Fly Larva, 14; Plant Lice and Scale Bugs, 15; Water Bugs and Beetles, 17; Getting Acquainted with the Library, 18.

CHAPTER II

STUDIES OF INSECTS 20

Field Studies, 20; Grasshoppers or Locusts, 23; Comparative Study of Orthoptera, 28; Key to Orthoptera, 29; Dragon Fly, 30; Honeybee, 31; General Study of Insects, 33; Review of Insects, 35; Key to Principal Orders, 36; Some Common Butterflies, 38; Summary of Insects, 37; General Review and Library Exercise, 40.

CHAPTER III

THE CONNECTION BETWEEN STRUCTURE AND FUNCTION 44

Protozoa: A Study of the Cell, 44; Comparative Study, 48; Review and Library Exercise, 49; Sponges, 51; Review and Library Exercise, 53. Coelenterates: Hydra, 54; Comparative Study, 58; Review and Library Exercise, 59. Worms: Living Earthworm, 61; External Morphology, 64; Internal Morphology, 64; Microscopic Anatomy, 66; Summary, 68; Review and Library Exercise, 70; Connection between Structure and Function, 72.

CHAPTER IV

ADAPTATION TO SURROUNDINGS 73

Crayfishes: Living Crayfish, 73; Morphology, 75; Summary, 79; Review and Library Exercise, 81.

CHAPTER V

ADAPTATION FOR PROTECTION FROM ENEMIES 83

Mollusca: Clam, 83; Snail, 87; Squid, 89; Comparative Study, 91; Review and Library Exercise, 92. Comparative Study of Exoskeletons, 93; Protective Coloration, 94; Animal Associations, 96; Protective Habits and Powers, 98; Defensive Structures, 99; Thesis: "Adaptation for Protection," 99.

CHAPTER VI

VERTEBRATES 101

Fishes: Living Fish, 101; External Structure, 103; Mouth and Gills, 105; Alimentary Canal and Circulatory System, 107; Review and Library Exercise, 110. Primitive Chordates, 112. Amphibia: Living Frog or Toad, 114; Mouth, 116; Organs of Digestion, Absorption, and Excretion, 117; Organs of Circulation and Respiration, 120; Nervous System, 123; Endoskeleton, 125; Comparative Study, 129; General Review and Library Exercise, 129. Reptiles: Living Snake, Lizard, and Turtle, 130; Review and Library Exercise, 132. Birds: Living Pigeon, 133; Plumage, 136; Birds and Migration, 139; Laboratory Exercise, 139; Field Work, 141; Review and Library Exercise, 142; Migration in General, 145. Mammals: Rodents: Domestic Rabbit, 146; Wild Rabbit, 148; Guinea Pig or White Rat, 149; Squirrel, 150; Library Exercise, 152. Carnivora: Laboratory Exercise, 153; Library Exercise, 155. Ungulates: Laboratory Exercise, 157; Library Exercise, 160. The Horse, 162. Homology of the Vertebrate Skeleton, 168.

CHAPTER VII

ADAPTATIONS FOR THE PRESERVATION OF THE SPECIES 170

Methods of Reproduction: Simple or Asexual Method, 170; Complex or Sexual Method, 171. Development: Structure of an Egg, 172; Development of an Egg, 173; Metamorphosis of a Mosquito, 174; Metamorphosis of a Butterfly, 176; Development of the Chick, 177. Protection and Care of Young: Library Exercise, 179. Adaptation for the Preservation of the Species: Review and Library Exercise, 180.

CHAPTER VIII

POULTRY 182

GLOSSARY 189

A GUIDE FOR THE STUDY OF ANIMALS

CHAPTER I

INTRODUCTORY STUDIES OF LIVING ANIMALS

In the following brief exercises the primary purpose is to arouse an active, attentive interest on the part of the pupil in various forms of animal life which may be at hand, reminding him of what and how various creatures eat, how they breathe, how they get ideas of the world, how they get about, and perhaps how they succeed where others fail. Of secondary importance is the introduction of laboratory methods by easy stages. The pupil should feel that his natural curiosity is only being directed to definite ends and that he is free to investigate in his own way.

The types here given are only a few of the many to be found in the early fall, and these exercises in several cases may be used for other forms than those definitely mentioned. There should be a great deal of promiscuous collecting by the class, and in the mass of material gathered the following types will probably be fairly abundant.

THE LIVING FLY

Materials.

Living flies in cages and individual specimens in small wide-mouth vials with cotton stoppers for the admission of air. Sugar crystals may be used for feeding. Simple lenses.

Observations.

Notice the division of the body into three regions: head, thorax, and abdomen. The six legs, the large wings, and the small feelers may be easily found, as are the large eyes, the extensible mouth, and beneath the larger wings the small undeveloped ones looking like tiny knobs.

1. State the general color of your specimen and give any special markings on its body.

2. Is the body smooth or has it a covering of any kind? Do you regard the fly as a cleanly animal? Why?

3. Under what conditions does the fly use its legs? its wings? What enables it to walk upside down? What use can you assign to the small wings?

4. Judging by the relative size of the feelers and the eyes, do you think the fly relies more upon its sight or its feeling? Since the eyes can probably see you any place where you see them, determine through how much of a circle the fly can see.

5. How does the fly eat? Does it eat solid or liquid food?

6. Where is the extensible mouth (proboscis) kept when not in use? What is the fly doing when "washing its face"?

7. From your own observation in the barn and the alley what do you know about the fly's cleanliness in choosing its food? How would it affect articles in the pantry?

8. From the foregoing statements show how flies may be a serious factor in dealing with disease.

9. What means may be employed as protection against adult flies? against their breeding places and "maggots"?

Suggested drawing.

a. The entire fly, seen from above, x 4.

THE LIVING MAGGOT

Materials.

Living specimens in pans or cotton-stoppered bottles, with some food material and moist cloth or paper; lenses.

Observations.

Notice the general worm-like form of the maggot, or grub, the plain and uniform color, and the absence of all elaborate structures, as wings and feelers.

1. Since this creature is destined to become a flying or walking insect, what organs will have to appear? Is there any indication of these structures at present?

2. Give the color of the specimen, and explain how the presence or absence of strong light seems to have affected the color. Is this effect usual in animals or plants that you know?

3. Tell how the animal gets from place to place, describing any special structures you find which aid in this work.

4. How can you tell the head end? Tell how the amount of work that the mouth and mouth parts do affects their size and indirectly that of the region where they are.

5. What senses and sense organs has the maggot? Test any of these senses or organs gently, by any means at your disposal, or recall any experience you have had along this line. Which senses or organs seem to be best developed?

6. Explain briefly how the active or sluggish habits either determine or are determined by the condition of the senses or sense organs.

7. Since "Mother Nature" seems to want maggots to develop rapidly, tell how she economizes in energy and material when forming them.

8. Show how the development of maggots in refuse matter is actually beneficial.

9. From the standpoint of flies and human welfare, show why maggots should not be allowed to live,--stating how they may be prevented.

10. Look up the story of the pupa of the house fly; the development and work of the botfly; of the ox-warble; of the tsetse fly.

Suggested drawing.

a. The maggot or grub, side view, x 4.

THE LIVING COCKROACH

Materials.

Individual specimens in cages, jars, or wide-mouth vials with cotton stoppers to admit air. Several roaches in large cages with material for food and concealment.

Observations.

1. What is the general color and the average size of cockroaches?

2. During what time of the day are roaches most active? Where do they hide at other times? How do their shape and color aid concealment? Note any odd or striking colors or marks which might make them distinguishable to their mates.

3. Is the roach a quick or a slow moving animal? How does it get about,--by running, jumping, walking, crawling, swimming, or flying? Turn your specimen on its back and see how it recovers its proper position. Notice the relative size and development of the wings and their use in flying.

4. If uninjured, your specimen has six legs. Why don't they step on each other? Notice the stiff hairs on the legs and the white pads under the feet. How would these structures be useful to the animal?

5. The large, black, shiny eyes are on the front and sides of the head; the long "horns," or feelers, are attached just below the eyes. Upon which sense, sight or feeling, do you think the roach depends more? Explain your statement.

6. Beside the mouth are a long and a short pair of "feelers"; perhaps these are for tasting or smelling. What do roaches like to eat? Do they choose their food? What damage do they do?

7. How can a house be rid of cockroaches?

Suggested drawing.

a. A cockroach, seen from above.

THE LIVING SPIDER

Materials.

Living spiders, preferably large ones, in cages; individual specimens in battery jars or wide-mouth bottles. Cocoons. Simple lenses.

Observations.

Each pupil may feel sure that if treated fairly any of the common spiders may be handled without fear of bite or injury.

1. Note that the spider's body is of two regions, the head-thorax and the abdomen, and that it is supported by eight legs. To what part of the body are the legs attached?

2. Find the feelers; if they are club-shaped, your specimen is a male. State their number and tell where they are attached. What is the sex of your spider?

3. Usually there are eight tiny near-sighted eyes on the front of the head. State the color of the eyes and by a diagram indicate their arrangement.

4. With what kind of material is the body covered (use the lens)?

5. What is the color of your specimen? What special markings has it?

Find out the name of your kind of spider.

6. Holding the spider aloft in your fingers, allow it to drop upon the thread it will spin, and watch it climb and spin. Record the number of the spinners, their situation, and how they act. Are the threads sticky? If so, why doesn't the

spider stick to its web? Is the web used for a home or for a snare?

7. Try to discover how the feet are enabled to cling to the thread.

8. Examine a cocoon, noting its outer form and structure, and look for an opening at the top. If you can open a cocoon carefully with scissors, look for its two coats and inspect its contents.

9. State three uses for the spider's silk.

10. What is the work of spiders amongst the animal population of the earth, or of what use are they?

11. Out of doors find webs of various kinds: wheel web, tent web, triangle web, etc.

12. How do the jumping spiders differ from others in their spinning and feeding habits?

13. Look up what is meant by ballooning spiders. Find out when ballooning occurs and what is accomplished by it.

Suggested drawings.

a. The entire spider, seen from above.

b. A cocoon.

THE LIVING CRICKET

Materials.

Living crickets in cages, with materials for food and concealment, and individual specimens in wide-mouth bottles or vials with cotton stoppers.

Observations.

1. What is the average size and the general color of crickets?

2. Just what do they do when you try to catch them? What structures enable them to do these things?

3. Of the three pairs of legs, which extend sidewise for running or grasping, and which backward for jumping or climbing? What structures have the legs to enable them to do their work properly?

4. Notice how well developed the cricket's wings are, and state how much they are used or how they influence the habits of the animal.

5. How many projecting spines are there on the hinder end of the body? Are they ornamental or useful? how? The female crickets have a special spear-shaped spine for depositing eggs.

6. In a column make a list of the senses (sight, feeling, etc.), and opposite each state what kind of an organ is used and where it is located. The ears are oblong white spots on the second long piece of the front legs.

7. Find out whether the cricket chews solid food or sucks liquid food, and whether it has biting jaws or protrusible lips. See whether it will attack a toothpick or your finger, and if the crickets have been confined long, whether there has been any attempt at cannabalism. Is its natural food animal or vegetable matter?

8. How do crickets chirp?

9. What work do they do in nature?

10. How does a baby cricket develop?

THE LIVING GRASSHOPPER OR LOCUST

Materials.

Individual specimens in wide-mouth bottles or jars, and other specimens in cages, with turf or foliage for food and concealment. Simple lenses.

Observations.

Notice the form and size of your specimen, its color, the number of its legs and of its feelers. Find the eyes; the two large eyes, a tiny one between the two feelers, and near the inner edge of each large eye, another tiny one. With a lens notice the markings on the large eyes. Find the mouth, and note its lips and finger-like feelers. Draw out an outer wing, and then carefully draw out the delicate under wing, allowing them both to fold into place again. Under the wings find the circular or crescent-shaped membranes, the eardrums. Watch the grasshopper's body expand and contract in breathing, and find the small breathing holes along each side the body. Compare its rate of breathing with your own.

Questions.

1. In what surroundings and how does the grasshopper's color protect it? What color markings has it which might serve for other grasshoppers to see as signals? Explain how this signaling is done.

2. Explain how the legs are placed so as to act as springs in jumping and alighting.

3. What advantages in having the wings attached on the upper side and the legs on the under side of the body?

4. Explain how the small wings are protective, and how the large ones are protected.

5. Why is it better for the grasshopper to have its mouth on the under side of its head instead of in front?

6. The large eyes are supposed to be far-sighted, the small ones near-sighted. State how the large eyes have the more advantageous position, and around how much of a circle they can see.

7. Describe how the grasshopper breathes.

THE LIVING BUTTERFLY OR MOTH

Materials.

Individual specimens in large jars or cages, and other specimens in cages with foliage; simple lenses and a needle or pin.

Observations.

Butterflies may generally be distinguished from moths by their habit of holding their wings together above them when at rest, by the feelers which are knobbed at the end, and by the rather slender abdomen. Moths generally either fold their wings or hold them outstretched, their feelers are not knobbed, and their bodies are rather bulky.

Observe these points in your specimen and the colors of the upper and under sides of the wings. Find the large eyes and examine them with a lens. With the needle or pin carefully uncoil the sucking tube which you may find under the head between two shields. Note the fuzziness of the body and the "dust" which covers the wings. Examine some of this dust under a lens.

Questions.

1. Is your specimen a butterfly or a moth? Prove your statement. If possible, give the name of your specimen.

2. Write a description of your specimen--its size, general color, and special color pattern.

3. Describe the sucking tube, or "proboscis," and name some flowers from which it might obtain nectar. Try to find out how the tube is operated.

4. Why is it that moths and butterflies never bite? Do they sting? How do you think they protect themselves from enemies?

5. State how the fuzz and dust on your specimen might influence a bird's liking for it.

6. Contrast the size and usefulness of the wings of the butterfly with those of

some other insect you know about; contrast their legs; state how development of one set of structures may cause another set to be simple or feeble.

7. Most moths are active by night. What explanation can you give for their large eyes and expanded feelers? Feelers of insects may be for any or all of the following: touch, taste, smell, and hearing.

Suggested drawings.

a. The butterfly or moth.

b. An antenna (feeler).

THE LIVING CATERPILLAR

Materials.

Living caterpillars in cages or covered jars for individual study, and other specimens in cages with foliage for food or concealment.

Observations.

The pupil should observe the general form and external construction of the caterpillar, watching it feeding, in action, and at rest.

Notice how the creature moves. Find its head, its segments (similar divisions of the body), and its breathing holes along the sides of the body. Try to find its eyes, any feelers, wings or paddles. Try to loosen it from its support; find the tiny hooks on the feet for clinging fast.

Questions.

1. Give the general color of your specimen and explain how this color may make it conspicuous or may aid its concealment.

2. Describe the outer surface or covering of the caterpillar. What structures, if any, are there, which might make the animal distasteful or inedible?

3. How many pairs of legs are there? How are they distributed along the body? Counting the segments, state which ones bear no legs.

4. To what extent do the legs act in locomotion? Are they mere organs for attachment while the body swings forward and backward, or do the legs do this, as in a horse? Make a complete statement.

5. Notice the openings of the internal breathing tubes. How are they protected against dust and other foreign matter?

6. Does the caterpillar seem to be a warm-blooded animal? State how the free access of air along the body would influence internal temperature.

7. What do you know about a caterpillar's appetite? How might caterpillars be beneficial or harmful? What means has nature of holding their numbers in check?

8. Recalling that caterpillars finally "sleep" for several days or weeks and awaken as winged creatures, how can you account for their appetites?

THE TUSSOCK MOTH

Materials.

Directions for the study of the caterpillar stage will be found in the exercise "The Living Caterpillar," and directions for the study of the adult male form will be found in the exercise "The Living Butterfly or Moth." The female tussock moth is a wingless, thick-bodied creature, gray in color, very downy, and about three fourths of an inch long. The following directions apply more particularly to the study of the cocoons and the general harmfulness of the tussock moth.

This exercise may be done best outside of the classroom, the pupil answering the questions on scrap paper and rewriting these notes in the laboratory. Living caterpillars, cocoons, some of them bearing their frothy masses of wax and eggs, pupae, and adult moths of both sexes may be used in the laboratory.

Observations and Questions.

1. On what kinds of trees are the cocoons and the caterpillars generally found? What effect have the caterpillars on the trees, and what may possibly be the final effect upon the trees of the locality or the entire district?

2. Upon what part of the tree are the cocoons made, and why? Where on the bark are they, and why?

3. Is the opening of the cocoon at the upper or the lower end? What reason can you assign for this?

4. Count the number of cocoons upon the entire tree or estimate it by counting those upon a part of the tree. Now count the number of eggs on a cocoon. Assuming that one half of the cocoons bear eggs, calculate the number of caterpillars on a tree next year.

5. How is the waxy covering of the eggs a particularly good protection against winter weather?

6. Investigate the interiors of several cocoons and state what you find.

7. On the pupa find the jointed and tapering hinder end, abdomen, and at the head region and lying along the under side, the marks of the legs and the feelers, and possibly the wings, all pressed close against the body. Find also the breathing pores along the sides of the abdomen.

8. Unlikeness between male and female is called "sexual dimorphism." Explain how the tussock moth shows this. For what work does each form seem particularly adapted?

9. What methods would you use that the tussock moth might be destroyed or kept out of a community?

The numerous small worm-like creatures often found are the caterpillar stages of another insect, an ichneumon fly, which laid its eggs under the skin of the tussock caterpillar. How has their development affected that of the tussock moth? What great result does nature accomplish by this arrangement?

Suggested drawings.

a. A caterpillar, x 2.

b. A cocoon with its egg mass.

c. A pupa as seen from the under side.

d. An adult moth, either male or female.

THE LIVING BEETLE

Materials.

Living beetles in cages, together with portions of the plant upon which they are found; or if water beetles are used, they should be kept in aquaria. Individual specimens in battery jars or wide-mouth bottles, and preserved beetles in pans or vials for reference.

Observations.

1. Upon what plant or in what surroundings is your kind of beetle generally found? If you can, give its common name.

2. What is the length, breadth, and thickness of your beetle? Would you describe it as a "small" insect or a "large" one?

3. Of what general color is it? Describe any color markings you see.

4. If any of the legs differ from the others or are of peculiar shape or length, describe them and tell what you think they may be fitted to do.

5. As a rule, beetles have harder "shells" than other insects. Does this shell completely inclose the body, or can you find any soft parts exposed? How are the head, thorax, and abdomen joined so as to carry out the apparent purpose of protection? What is the outline of the body--a continuous line or one with many irregularities?

6. If possible, try to lift up one side of the "shell" from the hinder end of the body. You will discover that this portion of the shell is a pair of hard sheath wings, and beneath them is another pair. How are the under ones unlike the upper in size and texture? in use? in arrangement when not in use?

7. Does a beetle spend most of its life on the wing, like bees and flies? How might the body covering and the structure of the outer wings affect or determine the beetle's habits, even against its will?

8. Are the feelers or are the eyes of your specimen more easily distinguished? Upon which of the special senses does it seem to place most dependence?

9. Is your kind of beetle good for anything, either in nature or in human affairs? Make a statement regarding what good or what harm it may do.

THE DAMSEL FLY LARVA

Materials.

Living larvae of the damsel fly in shallow watch glasses of water for individual use, and others in large pans or aquaria. Simple lenses or dissecting microscopes.

Observations.

1. What is the color and the shape of the larva? how long is it? Notice in what surroundings in the water the larva lives, and answer to yourself how its form and color would protect it in those surroundings.

2. Since the larva is an insect, though immature, its body is composed of three regions: head, thorax, and abdomen. How do these regions differ from each other?

3. What structures has your specimen to enable it to move from place to place? If fully developed wings are not present, what indication is there of their being formed?

4. What sense organs has the larva? Which ones seem to be the largest and

most useful?

5. Although the aquatic larva is preparing for adult life in the air, there should be some arrangement for securing air in the water. Where do you find outgrowths of the skin which might increase the air-absorbing surface? How many of these structures are there? Look within them for the air tubes,--fine branching black lines.

6. If possible, without injury to the specimen, examine the larva's mouth. Try to discover how it is used and how it is protected when not in use.

Suggested drawings.

a. The entire larva, x 4.

b. A gill, as seen through the microscope.

c. The feeding apparatus, x 10.

PLANT LICE AND SCALE BUGS

Plant Lice (Aphids)

Materials.

Plant lice on various kinds of plants, such as house plants, golden glow, and other plants from the garden or field. Garden asters with root lice (the asters should be transplanted into pots).

Observations.

1. Describe the size, appearance, and colors of the plant lice in your collection and their relation to the host plant.

2. Are the lice active or sluggish? (Compare with a house fly, for instance.) What proportion of them have wings? What is the usual method of locomotion?

3. Examining a single winged specimen, how many wings do you find? How

do they fold? What is the character of the wings?

4. What is the food of the plant lice? How is the food obtained? (With a hand lens identify the piercing organ.) On what parts of the plants are they found? Does the plant show any indication of being harmed by the lice? If so, how?

5. Make a count of the plant lice upon a portion of a plant and estimate the whole number upon a plant. Why are plant lice a very serious pest?

6. If any plant lice have ants associated with them, study the behavior of the ants in this curious relationship. What advantages result from this relationship of ant and aphis to either or both insects?

Scale Bugs

Materials.

Twigs of trees, leaves, fruit, ferns, etc., infested with these bugs. If possible, have samples of San Jose scales, maple scales, and oyster scales.

Observations.

1. What is the general size and appearance of the various scale bugs in your collection? How do they differ in form and size and color?

2. Remove a scale and study it carefully with a lens. What is under the scale? Of what is the scale composed? What do you discover about these bugs to indicate that they are really insects?

3. What can be said about the number of scale bugs? Why are they difficult to exterminate? How can they be distributed from one place to another, as from orchard to orchard, since only the males have wings?

Drawings suggested.

a. A single aphis as seen with hand lens.

b. Various scale bugs as seen with a hand lens.

c. Twigs showing the distribution and numbers of scale bugs.

WATER BUGS AND BEETLES

Water Bugs

Materials.

Water bugs and beetles of several species in small aquaria covered with a wire net.

Observations.

1. With what legs does the bug swim? Describe their appearance and tell how used and how fitted for this use. Which legs are not used in swimming?

2. Remove the bug from the water for a moment to test other methods of locomotion, as jumping, crawling, flying. What do you discover?

3. Watch the bug as it gets a fresh supply of air, and describe the process. Where is the air stored for use when under water? Does the bug sink or rise when it stops swimming? Why?

4. Identify the bug's mouth parts. What is their appearance and probable manner of use? How are the forelegs fitted for grasping food?

5. What is the shape of the body? What is the position of the wings? Do the two cover (fore) wings meet in a straight line or do they cross at their tips? Are they smooth throughout and sheath-like, or are they thick at the base and thin at the tips?

Summary.

Summarize your study of the bug by enumerating the various ways the bug is adapted for life in the water.

Water Beetles

Use the same questions for the study of water beetles as for water bugs. In addition answer the following:--

1. Identify the eyes of the whirligig beetle and note their peculiar construction. How can you explain this peculiar form of the eyes on the basis of use? Why are the antennae of both water bugs and beetles so small?

Suggested drawings.

a. The dorsal view of both bug and beetle.

b. Ventral view of the bug's head to show the beak and first pair of legs.

GETTING ACQUAINTED WITH THE LIBRARY

Directions.

The books in a zoological library may be roughly divided into three groups:--

1. Reference books. a. Advanced textbooks. b. Elementary textbooks. c. Natural histories. d. Books for classifying or naming animals. 2. Descriptive books. e. Life histories and habits of animals. f. Adventures with animals--popular accounts of animals seen on walks and travels. 3. Economic zoology. g. Books on harmful animals and methods of destroying them. h. Books on useful nondomesticated animals and their products. i. Books on domestic animals. j. Books of a general nature not included in the above.

Examine as many of the books in your library as you can and record for each one in your notebook:--

1. Title of the book; author's name; publisher; date of publication.

2. The kind of book as classified above.

3. What it includes or what animals or topics are covered by the book.

4. Whether the style is popular or technical, i.e. whether it is easy for you to

read.

5. The general character of its illustrations and whether they appear to be especially helpful.

6. Comments on the value or interest of the book as it appears to you.

7. Select a book which interests you, for future reading.

CHAPTER II

STUDIES OF INSECTS

The effect of great numbers upon the structure and habits of animals. The use of keys in finding the names of animals.

1. FIELD STUDIES

Materials.

1. Boxes for carrying insects. 2. A net. This may be homemade, using mosquito netting or fish net and a stout wire. If it is to be used for a dragnet for water insects, the wire must be stout and the netting strong. Make the net twice as long as wide. 3. A cyanide jar for killing insects. 4. A few paper triangles for carrying butterflies. 5. A notebook.

Note.--Your instructor will give directions for obtaining the material called for in 3, 4, and 5.

Directions.

Look carefully and quietly in the various situations noted below. Do not be in a hurry. Weedy meadows or vacant lots and neglected roadsides are good places for your first trips. Note concerning each insect found: (a) its name or something by which to identify it, (b) where you found it, (c) what it was doing, (d) its probable food. Record these observations in your notebook. Make a special study of such insects as your instructor may designate.

Where to look for Various Insects

Grasshoppers, locusts, katydids. Look along roadsides, waste places, gardens, especially weedy ones, weedy lots, and grassy meadows and pastures.

Crickets. Under old boards, along the edges of board or stone walks, along fences.

Beetles. Same locations as for crickets, and also on flowering plants, under loose bark of trees and stumps, in rotten logs, etc. For water beetles drag edges of ponds and streams.

Dragon flies. Along water-courses, ponds, and swamps. Drag ponds and ditches for larvae.

Bees. On flowering plants, especially on large patches of wild asters, golden-rods, and thistles.

Wasps. Sandy stretches,--especially along the water,--among flowering plants, under the eaves and roofs of outbuildings. Nests may be found in these latter places.

Butterflies and moths. In fields where there are many flowering plants; look carefully on the leaves of plants for caterpillars, and for eggs. Also look very carefully on the under side of leaves, on twigs, and on the bark of trees for chrysalids of butterflies and cocoons of moths.

Bugs. In same locations as for bees and grasshoppers and water beetles. Also on fruit.

Aphids. On the fresh growing tops of plants.

Tree hoppers. On trees and shrubs. Hold your net on the under side of branches and shake the branch vigorously.

Flies. Around decaying substances, as garbage, fruit, etc.; on flowering plants.

Ants. Sandy waste places, decayed logs, along walks, often in kitchens.

Note.--At night many kinds of insects fly around electric lights or into open windows, attracted by the light and may easily be collected.

Form for Field Trip Report

The notes taken on a field trip may be conveniently tabulated for permanent record in the form indicated below:--

FIELD TRIP REPORT

Date_____ Time_____ Locality_____ Pupil's Name_____ ---------------
--- Name of Insect | Where Found |
What it was Doing | Probable Food ---------------+-------------+-------------------
+-------------- | | | | | |

In case the name of the insect is not known to you, use a number and some designation as to color or other mark by which it may be known until you have leisure to look up its name by means of keys or books on insects.

Special Field Studies

The questions below may be used for a more careful field study of any insect.

1. Just where was the insect found?

2. Note carefully what the insects are doing before they are disturbed by your presence. What did the insects do when you disturbed them? If you think this related to securing safety, explain what leads you to think so.

3. What senses do you conclude are well developed? Reason for your conclusion.

4. Has the insect a home? If so, what is its character?

5. What is the color? What is the relation between the color of the insect and its surroundings?

6. Is the insect solitary in its habits or associated with others of the same species? If in association with others, note the numbers, and what they are doing.

7. What modes of locomotion do you observe in this insect? Which is the most common? If it flies or jumps, note the distance.

8. If you find the young, note whether they differ from the adult in general appearance, and if so, in what ways they differ. Do they differ in food?

9. What other insects do you find in the same habitat?

2. A STUDY OF GRASSHOPPERS (LOCUSTS)

Insects adapted to Life in Grassy Meadows and Fields

Materials.

Both living and dead specimens of grasshoppers. Various stages of young grasshoppers either dead or living. Some mounted specimens with wings spread. The wings of grasshoppers mounted in pairs between two glass slides for use with microscope or hand lens. Mounted preparations of mouth parts and tracheae.

Definitions.

Orthoptera, straight-winged insects, order to which belong grasshoppers, locusts, katydids, crickets, cockroaches, etc.

Vivarium, a cage in which living animals are kept.

Anterior, toward the head of an animal.

Posterior, opposite to anterior.

Dorsal, the upper surface of an animal.

Ventral, opposite to dorsal.

Regions, principal divisions of the body of an animal.

Head, thorax, and abdomen, the three distinct regions into which the body of a grasshopper is divided.

Somite, a ring-like division of the body of an animal.

Prothorax, mesothorax, and metathorax, the three divisions or somites into which the thorax of any insect is divided. A pair of legs is borne on each division.

Exoskeleton, an external skeleton.

Femur, tibia, and tarsus, the three principal divisions of the leg corresponding to thigh, shank, and foot.

Veins, thread-like thickenings of the wings.

Ocelli, the single or simple eyes of an insect, composed of a single eye element.

Compound eyes, made up of many eye elements.

Auditory sacs, organs for hearing in many animals.

Antennae, the feelers borne on the head.

Labrum, the upper lip.

Labium, the lower lip, formed by the growing together of the second maxillae.

Mandibles, primary jaws situated under the labrum.

Maxillae, secondary jaws just in front of the labium, each composed of three parts, a palp, a spoon, and a tooth.

Palps, the jointed finger-like structures used to handle food, one pair on the labium and one pair on the maxillae.

Spiracles, openings into the trachea found along the sides of the abdomen and thorax.

Tracheae, slender tubes used for breathing organs among insects. They carry the air direct to the tissues in all parts of the body.

Ovipositors, structures on the posterior end of the abdomen of a female, used to deposit eggs.

Metamorphosis, refers to the development of the young of animals when striking changes in structure occur in the course of their growth. Metamorphosis is called complete when the young have no resemblance to the adults, and incomplete when there is a resemblance to the adult. In complete metamorphosis the stages are larva, pupa, and adult. In incomplete metamorphosis the stages are nymph and adult.

Observations.

The Body.

1. Show how the shape of the grasshopper's body is well adapted to its needs.

2. Which region of the body is the thickest? What seems to be the reason for this? Which regions are capable of movement?

Locomotion.

3. What are the various kinds of locomotion a grasshopper can use? Which are used in the vivarium and which when free in the laboratory?

4. Which legs are used in jumping? How are these legs especially adapted to this, in length, structure and direction? Could a grasshopper jump if the third pair of legs were arranged like the other two pairs? Why?

5. How is the animal able to cling to grass stems and not slip down? What is the direction of the body in relation to the stem or grass blade?

6. What is the position of the wings when at rest? when in use? How do the hind wings fold? How are the principal veins of the wings arranged to permit or facilitate this folding?

7. Contrast the fore and hind wings with respect to thickness, size, and use.

8. To which somites of the thorax are the wings attached? Nearer which surface, the dorsal or ventral? Why?

Sense Organs.

9. Discover all you can about the uses of the antennae by carefully observing grasshoppers at rest, feeding, jumping and crawling, approaching an object or another grasshopper, etc.

10. How many compound eyes has the grasshopper? How many simple eyes? Where are they located?

Examine a preparation of the compound eye with the low power or as demonstrated with the stereopticon. What is the shape of an eye element of the compound eye? About how many eye elements are there in a compound eye?

Feeding.

11. Do grasshoppers eat and drink while in captivity? Put a fresh bunch of grass which has been sprinkled with water in a vivarium with grasshoppers that have had no food or drink for twenty-four hours and watch results.

12. What is the position of the grasshoppers in feeding? In what direction do the jaws move in feeding? Compare this with the direction of movement of your own jaws. What is the use of the palps? What do you think is the use of the "molasses" or saliva that flows from the mouth?

Respiration.

13. Describe the breathing movements of a grasshopper and explain the relation of the movements to inhalation and exhalation of air.

14. Find the exact location and number of spiracles on the abdomen. There are two pairs of spiracles on the thorax. Find them. How do the spiracles prevent the entrance of dust?

Describe a trachea as seen in a mounted preparation with the aid of a microscope or stereopticon.

Protection.

15. Explain how the colors of the grasshopper may be protective or useful when at rest in its natural habitat and when in flight.

16. Does the shell cover the entire body? What are the advantages of such a covering? A shell is likely to hinder activity, sensitiveness, and growth. How are such disadvantages overcome in this case?

17. What senses are probably most relied upon to detect approaching danger? Give evidence to support your answer.

18. What is the position of the hind legs when at rest? What relation has this to safety?

Reproduction and Development.

19. Describe the ovipositors and the probable method of their use. Describe the egg packets of grasshoppers, if discovered. About how many eggs in one? (They are sometimes seen against the glass sides of the vivaria.)

20. If you have young grasshoppers of various ages, arrange a set of them in what seems to you to be the order of their development. How do young grasshoppers differ from adults? What changes take place as they develop? What kind of metamorphosis is this?

Summary of Important Points in the Study of the Grasshopper

1. How many and what distinct regions of the body are there?

2. How many antennae? Compare their length with that of the body. What other sense organs did you discover?

3. How many legs? For what specially adapted? How?

4. How many wings? What is their resting position? How do the fore wings differ from the hind wings? How do the hind wings fold?

5. To what kind of feeding are they adapted, biting or sucking the food? How many and what sets of mouth parts are there?

6. How is air necessary for respiration obtained?

7. In what various ways are grasshoppers fitted for life in meadows and weed plots?

8. How do they meet winter conditions?

9. What kind of metamorphosis has the grasshopper?

Drawings suggested.

a. Side view with the legs and wings removed. Label all parts shown in this drawing. (See Definitions on pages 23 and 24 for names of parts.)

b. Face view of the head, showing the simple and compound eyes, the antennae, labrum, and palps.

c. One of the third pair of legs. Label parts.

d. A fore and a hind wing arranged in natural position.

e. A young grasshopper.

3. COMPARATIVE STUDY OF ORTHOPTERA

Materials.

Mounted specimens of various common species of orthoptera.

Observations.

1. Where does the insect live? What is its color?

2. What is the size and shape as compared with the grasshopper?

3. What is the length of the antennae as compared with the length of the body?

4. To what kind of locomotion are the legs adapted? How? Are the forelegs specially adapted for grasping?

5. What is the position of the wings when at rest? Are they large or small as compared with the size of the body?

6. Are the ovipositors long or short? (Compare with those of the grasshopper.)

7. Find the group to which the insect belongs and its name by the key in the following section.

4. KEY TO SOME COMMON ORTHOPTERA

A. Groups

GROUPS	LEGS	ANTENNAE	OTHER CHARACTERS
Cockroaches (Blattidae)	Similar, fitted for running	Long	Body flattened, wings folded on dorsal surface of the abdomen
Mantis (Mantidae)	First pair of legs enlarged for grasping	Rather long	Prothorax long and slender, wings folded on dorsal surface of abdomen
Walking stick (Phasmidae)	Similar, fitted for walking	Long	Body usually greatly elongated and stick-like, usually no wings
Short-horned	Hind legs fitted for jumping	Short	Body somewhat compressed,

wings folded on side of | grasshoppers | | abdomen | (Acrididae) |-------+--------
-------------------------+----------------- | | Body compressed, wings folded | Long-
horned | Long | on sides, tarsus four-jointed | grasshoppers | | | (Locustidae) |---
----+--------------------------------+----------------- | | Body somewhat flattened,
wings | Crickets | Long | folded on the back, tarsus | (Gryllidae) | | three-jointed
|

===
=======================

B. Species or Genera

===
==================== CHARACTERS OF SPECIES | COMMON
NAME | GROUPS -------------------------------------+----------------------+---------
------ Large size, brown color | American cockroach | Small size, pale brown |
"Croton bug" | Cockroaches Dark color, often wingless | Oriental cockroach | -
--------------------------------+----------------------+--------------- Body long,
anterior portion slender | Mantis or rear horse | Mantis ----------------------------
--------+----------------------+--------------- Long body, long legs, no wings |
Walking stick | Walking sticks -------------------------------------+-----------------
----+--------------- Very large size, wings very small | Lubber grasshopper |
Small to medium size, legs marked | Red-legged | with red | grasshopper |
Short-horned Large size, greenish brown color | Differential locust |
grasshoppers Medium to large size, sand color | Carolina locusts | (gray) | | ----
--------------------------------+----------------------+--------------- Rather large,
green, wings large | Angle-wing katydid | and angled | | Long-horned Small to
rather large, usually green | Meadow grasshopper | grasshoppers Wingless,
brown color | Cricket grasshopper | -------------------------------------+-------------
---------+--------------- Usually rather large, black | Field cricket | Crickets
Wingless,   front   legs   shovel-shaped | Mole   cricket |
===
=====================

5. THE DRAGON FLY

An Insect adapted to Aerial Life

Materials.

Mounted specimens of dragon flies, some moist preserved specimens, living specimens if practicable, simple lenses.

Observations.

1. Identify the three regions of the body and note the presence of a distinct neck. What is the length of the insect? What is its general form? If you have living specimens, discover what movements the head and abdomen are capable of making.

2. What is the position and general character of the wings? Explain how these wings are made very efficient for flying. Why should they not fold?

3. For what do the legs seem best adapted? Why?

4. Note the size of the eyes and of the antennae? How do you account for the great size of the eyes and the relatively small antennae?

5. What is the type of mouth parts, biting or sucking? If you have living dragon flies, try feeding them flies or mosquitoes and note how they are seized.

6. The food of dragon flies is mosquitoes and flies caught while on the wing. In what various ways is the dragon fly specialized for getting food in this manner?

Summary.

How is the dragon fly fitted for its aerial life with respect to its body, means and method of locomotion, sense organs, kind of food and manner of obtaining it?

Suggested drawing.

a. Dorsal view, showing veining of one wing.

6. THE HONEYBEE

A Study of Adaptations for Community Life

Materials.

Preserved specimens of workers in small vials and in watch glasses, and some mounted specimens. A demonstration case showing the three kinds of members of the community, stages in the development of the workers and queens and the cells in which they are reared, specimens of the comb. Small pieces of beeswax, a box of honey, and specimens of comb free for examination. Mounted preparations of mouth parts and stings. Simple lenses and compound microscopes.

Observations.

The Worker Bee.

1. Observe and describe the form, size, regions, and covering of the bee. What are its colors?

2. Observe and describe the texture, veining, relative size, and position of the wings. Discover how the fore and hind wings are hooked together. What advantage in having them hooked together?

3. For what kind of locomotion are the legs best adapted?

4. Find the pollen basket on the tarsus of a hind leg. How is it fitted for carrying pollen? What are the wax shears?

a. Examine and describe the structure at the posterior end of the body used for stinging. (Use a mounted preparation for this with low-power of microscope.)

b. The mouth parts are fitted for both biting and sucking. Find what makes this possible. (Use mounted preparation.)

5. Describe the antennae and the number, position, and shape of the eyes. Are the eyes fitted for keen sight? Give reason for answer.

6. The worker bee gathers honey and pollen and defends the entire community from enemies. What various adaptations fit it for this work?

The Community of Bees.

7. How do the workers, drones, and queen differ in general appearance?

8. Describe the appearance of the comb and the arrangement and shape of the cells. Why this shape? How are the cells closed when full of honey?

9. How do the cells used for rearing worker bees differ from those used for rearing queens? What is the appearance of the larvae? Of the pupa?

10. Examine and test in various ways a small piece of beeswax. What are the qualities possessed by this wax which make it suitable for making comb and protecting the home from storms?

Supplementary Studies of Bees

Materials.

For this study an observation hive of bees or opportunity to visit an apiary will be helpful. If neither are practicable, then look up the answers in books. There are government bulletins on bee-keeping and much helpful information can be obtained from large dealers in bees and bee supplies.

Observations.

1. How do bees protect their hives from rain and storm and light?

2. What are honey boxes? Where are they placed in the hive? Can the honey be removed late in the fall?

3. How is it safe to approach and handle bees in removing honey and caring for them?

4. What are their habits in entering and leaving the hive? What is the

appearance of a returning loaded worker bee?

5. How do bees survive the winter? Why are the drones driven away or killed?

6. Watch bees gathering nectar and pollen from flowers and describe the process. Try following a bee on its journeys.

7. When the bees are in the hive, how may you know the queen and drones from the workers?

8. What is swarming? When does it take place? How is the swarm hived?

9. What is the home of wild honeybees? How found?

Summary of the Study of Honeybees

How is the work of the community of bees divided among the bees? How is each fitted for the work? What do you think of the success of this kind of life? Give reasons for your answer.

7. GENERAL STUDY OF INSECTS[1]

[1] This study is intended for an alternative study in case it is not practicable to use the studies of living insects. With slight adaptation it can be made useful for any insect, either preserved or living.

Materials.

Both living and preserved specimens of the insects studied should be at hand, if practicable. There also should be specimens of the young.

Observations.

The Body.

1. What is the shape and size of the insect and the number of regions in its body? Does the shape seem to be in any way adapted to the mode of life of the insect? If so, how?

Locomotion.

2. What methods of locomotion has the insect? Which is the most used?

3. What is the position of the wings when at rest? What is the texture (e.g. thick, smooth, leathery, shell-like, membranous) of the fore and hind wings?

4. For what kind of locomotion are the legs fitted? How?

Sense Organs.

5. How many antennae has the insect? What is their character as to shape and length? How many simple and compound eyes?

Feeding.

6. What is the food of the insect? How are the mouth parts specially adapted to obtaining this food?

Note.--The mouth parts of insects may be jaws for biting, or may form a tube for sucking, or a beak for piercing and sucking.

Respiration.

7. Look for movements of the body indicating breathing, and describe what you find. Discover the location of the spiracles.

Protection.

8. What are the enemies of this insect? (Among the most important enemies of insects are birds, certain other insects, and various small vertebrates such as frogs, snakes, lizards, turtles, etc.) How does the insect protect itself from these enemies?

9. Describe the shell with respect to thickness and flexibility. What is the character of the surface as to roughness or smoothness or covering of hairs or scales?

Reproduction and Development.

Note.--It may be necessary to get answers to these questions from books.

10. Where are the eggs deposited? What is the number of the eggs? How soon do they hatch?

11. What is the food of the larva or nymph? Are the food habits of the insects harmful to man? If so, how?

12. Describe the larva as to form, color, and appendages. Is it capable of locomotion?

13. Is the metamorphosis complete or incomplete? If complete, describe the pupa and tell where it may be found.

Drawings.

There should be one drawing of the insect to show its general characteristics; usually a dorsal view is best. For other drawings ask your instructor.

8. A REVIEW OF INSECTS

Directions.

The answers to questions in this study may be conveniently written in the form of a table. Construct this table by placing the topics at the left and the names of insects at the top. Allow ample space, about one half inch for the horizontal spaces and one and one half inches in width for the vertical columns. Use one or two insects from each of the principal orders, letting the table extend across two opposite pages.

Topics.

1. What is the habitat?

2. What regions has the body?

3. How many antennae? What is their form?

4. What kinds of eyes has the insect? How many of each kind?

5. How many legs?

6. For what kind of locomotion are the legs adapted? Which legs are thus used?

7. How many wings? Membranous or thickened?

8. What is the position of the wings when at rest?

9. If the fore wings are thickened, what is their texture,--leathery, smooth and sheath-like, partly membranous, covered with scales?

10. What kind of mouth parts,--jaws for biting, a beak for piercing, a tube for sucking, adapted for both sucking and biting?

11. By what means is respiration accomplished?

Summary of Important Points from the Table

1. What characters are common to all the insects described in the table?

2. What are the various types of wings? Why do they vary?

3. What are the various types of legs? How are they characterized?

4. What are the various types of mouth parts?

5. Show how the variations in insects are related to the habitat and mode of life of the insect.

9. KEY TO THE PRINCIPAL ORDERS OF INSECTS

A^1 Insects with no wings. (See list below.)

A^2 Insects with wings B

B^1 With two pairs of wings. (See Note 1 below.) C

B^2 With one pair of wings Diptera

C^1 Both pairs of wings alike in structure, either membranous or scaly D

C^2 Fore and hind wings unlike in texture, fore wings fold over hind wings E

D^1 Both pairs of wings membranous, not covered with scales F

D^2 Both pairs of wings covered with scales; mouth parts tubular for sucking Lepidoptera

E^1 Fore wings very smooth, sheath or shell-like, meeting in a straight line when folded; legs adapted for walking, running, or swimming; mouth parts for biting Coleoptera

E^2 Wings not as in E^1 I

F^1 Wings membranous, usually folded or partly folded; few nerves G

F^2 Both pairs of membranous wings usually outspread, many nerves; mouth parts for biting H

G^1 Wings membranous, hooked together and partly folded, or outspread, few nerves in the wings; mouth parts for both biting and sucking; regions of the body usually very distinct Hymenoptera

G^2 Wings membranous, usually folded, few nerves; mouth parts, a beak for sucking and piercing Hemiptera

H^1 Outspread membranous wings, nearly equal in size; antennae very short and inconspicuous Odonata

H^2 As in F^2, but antennae not short; wings sometimes folded Neuroptera

H^3 Both pairs of wings membranous, folded above the back; fore wings much larger than hind wings; ovipositors long; mouth parts rudimentary Ephemerida

I^1 Fore wings folded over hind wings, crossing at their tips, which are membranous, base of wings thickened, mouth parts a beak for piercing Hemiptera

I^2 Fore wings leathery, folding either at side of body or on the back; mouth parts for biting, legs often adapted for jumping Orthoptera

Note 1.--When wings are folded, it will be helpful to remember that thickened fore or cover wings always have membranous wings folded beneath them.

Insects with no wings Order

a. Body long and slender, stick-like; legs for walking. Walking stick Orthoptera

b. Grasshopper-like. Cricket grasshopper Orthoptera

c. Small size; regions very distinct; abdomen spindle-shape. Ants Hymenoptera

d. Small size; ant-like in appearance; pale white. White ants Isoptera

e. Flattened body, small size; no compound eyes. Springtails and fish moths Thysanura

10. SUMMARY OF THE STUDIES OF INSECTS

The Effect of Great Numbers

1. Take some insect for illustration, as the house fly, mosquito, tussock moth, or aphis, and show how insects increase in numbers with great rapidity.

2. What can be said about the number of species of insects?

3. There is said to be great competition among insects. Why? For what?

4. How is the great increase of insects held in check by natural means?

5. What are the various habitats of insects? Give as many as you can with examples of insects that use the habitat.

6. Give examples to show how greatly the food of insects and the method of obtaining it varies.

7. Give some illustrations of the great muscular development of insects. Why is this needed?

8. In what various ways are insects protected against their enemies? Give examples to illustrate your statement.

9. Show how and why the great numbers of insects have affected the structure and mode of life of the insects.

Classification

1. By means of illustrations from your studies of insects show how classification is based upon likeness of structure.

2. In the same manner show how differences in structure affect classification.

3. Show how variation in the wings and mouth parts is used to separate insects into orders.

4. What are the principles of classification?

11. REVIEW AND LIBRARY EXERCISE ON INSECTS

General Topics

1. General characteristics of insects.

2. Principal orders of insects with characteristics and examples of each order.

3. Respiration and air sacs of insects. Use of air sacs in flight.

4. The heart and blood of insects. How the function of the blood differs from that of other animals, as man.

5. Special senses of insects: their character, location, and efficiency.

6. Sound-making organs of insects.

7. Power of communication among insects, as among ants, for example.

8. Organs for depositing eggs, ovipositors. How they vary.

9. Homes of insects. Evidences of architecture in some of the homes.

10. How some plants make homes for insects. Galls and gall insects.

11. In what various ways do insects survive the winter? Illustrate with examples.

12. Community life among insects. Types of communities.

13. Pollination of flowers by insects. Why insects do this work and how the flowers compel them to do it in the right manner. Value to the plants. Types of insects useful for this purpose.

14. Adaptations for protection against enemies. Classify these adaptations and illustrate with examples.

15. The principal insect pests of the orchard and their work.

16. The principal insect pests of the garden and the work of each.

17. The principal insect pests of shade trees and their characteristics.

18. The principal insect pests of the household and methods of extermination.

19. The work of birds in helping to keep the number of harmful insects down.

20. A spraying table showing what poisons are used, when and for what plants and insects.

21. The principal beneficial insects and the ways in which they are beneficial.

Special Topics

Much of the information called for by the topics below may be obtained from United States and state government bulletins. Most of these may be obtained free from the Department of Agriculture and from various state agricultural colleges, while others may be obtained by purchase at a nominal price.

Orthoptera.

1. Locust migrations and their cause.

2. The locust plagues of the "great plains."

3. Crickets and their "songs."

Hemiptera.

4. The fight against the orange scale of California.

5. History of the introduction and spread of the San Jose scale bug and the efforts to find a natural enemy. How people fight the pest.

6. Aphids.

7. Relations of ants and aphids.

8. Phylloxera and its work.

9. The methods of fighting the chinch bug.

10. Scale bugs.

11. Cochineal bug and the lacs.

Coleoptera.

12. The carrion beetle and its peculiar habits.

13. Fireflies.

14. Egyptian scarabs.

15. The curculio and methods of fighting it.

16. The weevils and their work.

17. History of the Colorado potato beetle.

18. Lady-bird beetles, their habits and use in exterminating harmful pests.

Diptera.

19. The investigations in Cuba of the cause of yellow fever.

20. The fight against yellow fever in New Orleans.

21. Methods of preventing plagues of mosquitoes.

22. How flies are carriers of disease. Methods of preventing plagues of flies.

23. The tsetse fly.

24. Sleeping sickness.

25. The house fly and typhoid.

26. Parasitic larvae of flies.

Lepidoptera.

27. The silkworm and the silk industry.

28. Story of the gypsy moth.

29. Life history of the clothes moth.

30. Harmful butterflies.

31. The tussock moth and its history.

32. Blastophaga and fig culture.

33. The codling moth and its work.

34. Cutworms.

35. The brown-tail moth.

Hymenoptera.

36. The honeybee and honey making.

37. Gall and gall insects.

38. The habits of the digger wasp.

39. The homes of ants. Habits of ants.

40. Slavery among ants.

41. Agricultural ants.

42. Homes of bees.

43. Ichneumon flies and their beneficial habit.

44. Evidences of intelligence among ants.

SOME COMMON BUTTERFLIES--A Reference Table and Key

===
======================== GROUP

|COMMON NAME |WING EXPANSE|BROODS |FOOD PLANTS OF
|HAUNTS OF THE | | IN INCHES | | CATERPILLAR | BUTTERFLY
|CHARACTERISTIC COLORS, MARKINGS, ETC.

--- Milkweed
Butterflies

|Monarch |4--4-1/2 |May and Oct|Milkweed and |Open fields | | | | dogbane |
everywhere |Brick-red color, veins black, borders of wings black

--- Fritillaries
or Silver Spots

|Variegated |1-3/4--2-1/2|August |Passion flower |Low fields | fritillary | | | |
|Orange-brown color, checkered with black, no silver spots. A southern |
species

|Regal fritillary |3--4 |July,Aug |Violets, pansies |Low fields |Upper side of
wings reddish with wavy black lines, hind wing dark

|Great spangled |3--4 |July,Aug |Violets, pansies |Meadows | fritillary | | | |
|Similar to idalia, but hind wings lighter. Silver spots on under | surface of
wings

|Silver-bordered |1-1/2 |Jul,Aug,Sep|Violets, pansies |Meadows, | fritillary | | |
| hillsides |Edge of wings tipped with silver, silver spots below

|Meadow fritillary|1-3/4 |Jul,Aug,Sep|Violets, pansies |Meadows |No silver
border, silver below

--- Checker
Spots

|Baltimore |1-1/4--2-1/2|June,July |Turtlehead and |Swamps | | | | aster |
|Groundwork of black with many red and white spots. Conspicuous border of|
red spots

|Harris checker |1-1/2 |June |Aster and daisy |Clover meadows | spot | | | |
|Wings dark bordered, lighter band across middle of wings

--- Crescent
Spots

|Silver crescent |1-1/4--2 |July |Asters |Roadsides |Groundwork of orange-red
mottled with black, silver crescents on under | margin of hind wings

|Pearl crescent |1-1/4--1-5/8|July,Sep |Asters, daisy |Roadsides |Similar to
silver crescent but colors are paler

--- Angle
wings

|Comma |2 |May,Jun,Aug|Elm, nettle, hop |Along woods | | | | | and waste | | | |
| places |Pale red, angled wings, under surface light gray marked with silver
commas

|Interrogation |2-1/2 |May,Jul,Aug|Elm, nettle, hop |Near trees |Similar to
comma, but marked with silver semi-colons

--- Tortoise
Shells

|Compton's |2-3/4 |Feb,Oct |Willow |Near water | tortoise | | | | |Looks much
like the angle wings, but has no silver spots

|Milberts's |1-3/4 |May,June, |Nettle |Roadsides | tortoise | | Aug,Sep | | |Broad,
reddish yellow band across both wings

|Mourning cloak |3 |Apr,Jul,Sep|Willow, poplar |Everywhere |Black with yellow or cream-bordered wings

--   The Beauties

|Red admiral |2 |May,Jul,Sep|Nettle, elm |Waste land |Bright red band circling across both wings

|Painted beauty |2 |May,Jul,Sep|Everlasting, |Thistles | | | | thistle, burdock| |Mottled with pink, black and white, under surface mottled, two large spots | on under surface of hind wing

|Thistle butterfly|2--2-1/4 |May,Jul,Sep|Thistles |Pastures |Like the painted beauty, but has several small eye spots

-- The White Admirals

|Red-spotted |3 |July |Wild cherry, |Near trees | purple | | | apple, etc. | |Purple and blue above, six red spots on under surface of wings

|Banded purple |2-1/2 |July |Hawthorn |Open woods |A broad white band across both wings

|Viceroy |2-1/2 |June,Aug |Poplar, willow |Roadsides |Imitates the monarch, but is smaller and has a black line across the hind | wings

-- The Satyrs

|Grass nymph |1-3/4 |July |Grass |Meadows |Dull brown, twenty spots in two rows across the wings

|Little wood satyr|1-3/4 |July |Grass |Hillsides |Dull brown, six spots

|Wood nymph |2 |July |Grass |Hillsides |Dull brown, two eye spots on each fore wing in a larger yellow spot

-- Hairstreaks

|Hop hairstreak |1-1/8 |May,July |Hop |About | | | | | shrubbery |Dark color, hind wings have slender tail-like projection and black spots | crowned with crimson

--- The Coppers

|American copper |1 |May,Jun,Sep|Sorrel |Everywhere |Orange-red fore wings spotted with black, hind wing with orange border

--- The Blues

|Common blue |1 |May,July |Pea |Roadsides |Male light violet, female lighter with reddish bordered wings

|Tailed blue |1 |May,Aug,Sep|Clover, etc. |Roadsides, | | | | | fields |Purplish violet color, has small tail-like projection on hind wings

--- The Whites

|Common white |2 |May,Jul,Sep|Mustard family |Gardens |White checkered with black on fore wings, female brownish

|Cabbage butterfly|2 |May,Jul,Sep|Cabbage, etc. |Gardens |White, black tip on fore wing, one or two spots on hind wing

--- The Sulphurs

|Common sulphur |m. 1-3/4, |May,Jun,Sep|Clover |Meadows | | f. 2-1/4 | | | |Yellow, bordered with black

|Cloudless sulphur|2-1/2 |July |Cassia and legumes|Fields |Canary-yellow color

Swallowtails

|Tiger swallowtail|3--5 |June,Aug |Cherry, tulip tree|Open woods |Yellow with black lines across wings

|Black swallowtail|3--4 |June,Aug |Parsley |Gardens, | | | | | roadsides |Black with two bands of yellow spots and one band of blue spots

|Green-clouded |3-3/4--4-3/4|June,Sep |Spice bush, |Open woods | swallowtail | | | sassafras | |Black, one row yellow spots, hind wing clouded with green

|Blue swallowtail |3-3/4--4-1/4|July,Sep |Dutchman's pipe |Near houses | | | | vine | |Black shaded with blue green, one row whitish spots

=== ========================

CHAPTER III

THE CONNECTION BETWEEN STRUCTURE AND FUNCTION

1. A STUDY OF THE CELL AND OF PROTOZOA

To show what Single Cells can Do

Materials.

Some single cells of plant or animal tissue, stained to show structure. Slides of a one-celled animal, stained. Living one-celled animals.

Definitions.

Cell, the smallest living unit.

Protoplasm, the living material composing the cell.

Nucleus, a dense bit of protoplasm, usually near the center of the cell, often staining dark.

Cytoplasm, the less dense protoplasm outside of the nucleus, usually taking a lighter stain.

Nucleolus, paranucleus or micronucleus, a very small, dense, dark-staining body, either within the nucleus (nucleolus) or near it (paranucleus or micronucleus)

Cell wall, the lifeless membrane surrounding many cells, secreted by the protoplasm.

Food balls, bits of food inside the cells of many one-celled animals, usually showing through the walls.

Food vacuole, a small drop of water containing digestive material and a food ball.

Contracting or pulsating vacuoles, small, clear spots in the cell, filled with water. In the living cell these disappear at intervals and then appear again.

Oral groove, a funnel-shaped groove in one side of some one-celled animals, conducting food to the mouth. In paramecium it often shows as an oblique line when the animal rolls.

Gullet, the inner end of the oral groove.

Cilia, numerous minute, vibrating, protoplasmic hairs on the surface of many cells.

Respiration, the passage of oxygen into the tissues of a living organism and of carbon dioxide out of them. These gases can pass through any thin, moist, organic membrane. When such a membrane separates two fluids which differ in the amount of oxygen they contain, oxygen passes to the fluid containing the smaller amount.[2] The same is true of carbon dioxide. Respiration is believed to occur in all living organisms.

[2] This passage of fluids through membranes is known as osmosis.

Digestion, the process of making food materials soluble, so that they can pass through membranes and be used to build up protoplasm. A few forms of cells are able to take in solid food and digest it in their protoplasm, but most cells can admit only fluid food.

Fission, a method of reproduction used in all cells, by which a cell divides itself into two, usually through the center. In some one-celled animals this may be preceded by conjugation, when two animals unite temporarily and exchange nuclear substance; or in some forms two cells may fuse and the resulting cell may divide. Buddingis a form of fission in which a small projection is formed on the parent cell and then cut off, making a new individual.

Protozoa (first animals), animals of one cell, existing alone or in loose colonies.

Observations.

1. Examine a single cell, stained to show structure. Identify the nucleus, cytoplasm, and, if present, the nucleolus or the micronucleus, and the cell wall. Draw to show the form of the cell and the details of its structure. Label all details.

2. Examine some stained paramecia. Select a typical one and identify in it nucleus, micronucleus, cytoplasm, and cell wall or cell membrane. You may also be able to see vacuoles, looking like holes in the stained protoplasm. Give reasons for considering this animal to be a single cell. Draw one, to show its cellular structure. Label all details.

3. Clean a slide and cover glass, place a drop of water containing living paramecia on the slide, cover it, and examine. What structures do you see which you saw in the stained paramecia? What structures do not show? Identify any new structures you may observe. Identify also the leading end and the side containing the oral groove.

4. Describe the shape of the animal.

What is the actual length of the animal?

5. After watching the animal for some time, describe the path followed by a given specimen as it crosses the field of the microscope. What reason can you see, if any, why this paramecium is moving? What external factors, if any, seem to determine the path it follows?

6. How rapidly do paramecia really move? What structures do they use in locomotion?

How do they manage to move in one direction, instead of alternately backward and forward? How do they manage to move in a straight line, though their bodies are not symmetrical?

7. What is the food of the paramecia? How do they find it? Find a specimen at rest and watch the oral groove. Suggest a method by which food may be collected into it. If possible, note the process of swallowing, and the resulting food ball.

Note.--If powdered carmine be placed in the water with some paramecia, it can be seen in the food balls a half hour or so later.

8. Where are the food balls located? Watch them in an individual until you notice their motion. Where are the larger food balls? the smaller ones? Assuming them to have been of approximately equal sizes when they were taken in, how can you account for differences now?

9. Where are the contracting vacuoles? How many are there? How often does one contract?

What is their function?

10. As you have been studying paramecia, to what external influences (as contact, heat, light, etc.) have you seen them respond? How do they show it when they do respond? Is such a response an advantage to them or not? What would be the result if they were not able to detect changes in their

surroundings?

11. Where does respiration occur in paramecia? Where do they obtain their supply of oxygen?

12. Among the paramecia you are studying you usually find at least one in the process of fission. Watch it until the halves separate, if you can. Compare the halves. Do they rank as parent and offspring? If so, which is which? If not, which are they, parent or offspring?

13. If you happen to find a pair conjugating, notice the process, as far as you can, in the living animals.

Suggested drawings.

a. A drawing to show all the details seen in the living paramecium.

b. A diagram to show the path followed by a paramecium to get around some obstacle.

c. Drawings to show that paramecia are constant in shape and yet flexible.

d. A drawing to show at least one stage in fission. This may be from a permanent preparation.

e. A drawing to show paramecia conjugating. This also may be from a permanent preparation.

f. Instead of all these separate drawings they may be combined into one. Represent the field of the microscope, and in it draw all necessary figures, to show the facts called for in the first five drawings and any other facts you have observed about living protozoa. Make the whole drawing to scale.

Summary of Important Points in the Study of Paramecia

1. Look back over your study of paramecia and list the different kinds of work you saw paramecia doing; also the kinds of work you infer they can do. What organs have they to use? When there is no organ to do a given thing, e.g.

to digest food, how is the work done?

2. What conditions are favorable to paramecia? Why are they so numerous under favorable conditions?

3. What would you call a successful animal? Are paramecia successful? Give reasons why they are or are not.

Comparative Study of Protozoa

To enlarge your idea of what a cell can do, spend as much more time on the one-celled animals as your course will permit. Any stagnant water may furnish several kinds. By means of reference books, identify as many as you can. In each case notice:--

1. Its size, shape and general appearance, comparing and contrasting it with paramecium.

2. Its usual surroundings, i.e. the conditions it has to meet.

3. The means it has of finding out facts about its surroundings.

4. The means it has of adjusting itself to its surroundings. For example, is it stationary? If so, what does it do when conditions change? Is it locomotory? If so, how effective is its locomotion?

5. What is its food? How does it find food?

6. Can it do as many kinds of work as paramecium can? Can it do any that paramecium cannot do? If so, what?

Review and Library Questions on Protozoa

1. What are the characteristics which distinguish protozoa from other animals?

2. What are the classes of protozoa? Characteristics of each class?

3. What is digestion? Where does it take place in the protozoa?

4. What results from the fact that the amoeba has no cell wall? (Give at least two points.)

5. In what ways are paramecia more specialized than amoeba are? How does their greater specialization show in their work?

6. What different methods of locomotion are shown among protozoa? By what means is locomotion accomplished in each case?

7. What is encysting? Name some protozoa which encyst. How long may an encysted animal live? When do they encyst? Why?

8. Give methods of reproduction among protozoa. Which method is fitted for rapid multiplication, for withstanding drouth; for renewing vitality?

9. Many scientists speak of protozoa as immortal. What argument is there to support such a statement?

10. Why are no protozoa large animals? Give at least two reasons.

11. Why are protozoa so numerous? Why more numerous in stagnant water?

12. Where are protozoa found?

13. Why are protozoa so widely distributed?

14. Write the probable history of a piece of chalk.

15. What connection is there between protozoa and some polishing powders?

16. Where in the human body are malarial protozoa found? How are they transferred from one human being to another? Why is there likely to be more malaria in newly settled regions than in older ones? If you were obliged to spend some time in a region where malaria existed, what precautions would you take?

17. Name other diseases caused by protozoa. How are they fought?

18. What beneficial effect have some protozoa upon the water of stagnant ponds and ditches? How may some forms injure water for household purposes?

19. Give at least three reasons for thinking that protozoa are the most ancient animals.

20. Why are protozoa of great importance to the world?

2. A STUDY OF SPONGES

To show how cells loosely associated may work together.

Materials.

The simplest of the many-celled animals are the sponges, which, with one exception, are salt-water forms. That one, the spongilla, is not easily found and is very difficult to maintain in the laboratory. For these reasons the material for this study is very meager, except at the seashore, and much of the work must be done from diagrams and reference books. Small simple preserved sponges and complex toilet sponge skeletons will also be used.

Definitions.

Body wall, the outer wall in bodies of the many-celled animals.

Central cavity, the cavity surrounded by the body wall in the simpler many-celled animals, as in the sponges.

Canals, channels through the body walls of sponges.

Inhalent pores, the outer ends of the canals.

Ostia, the inner ends of the canals.

Osculum, the large opening of the central cavity, at the distal end of the sponge.

Spicules, tiny needles of mineral substance found in the walls of many sponges.

Fibers, flexible threads of horny material found in the walls of many sponges.

Endoderm cells, cells lining the canals. They have cilia or flagella (projections larger than cilia).

Ectoderm cells, cells covering the outside of sponges and some other animals. In sponges it is believed that endoderm and ectoderm cells are able to exchange positions and functions.

Mesoglea, a jelly-like layer between the endoderm and ectoderm layers. In the sponges this contains many wandering cells, probably from the other layers.

Porifera (pore bearers), animals with many more or less independent cells, supported by solid skeletal parts and penetrated by a system of canals which open on the surface as pores.

Directions.

Study a simple sponge to see the shape, size, and point of attachment. Identify the osculum. In a diagram of a long section of a simple sponge identify the central cavity, body walls, canals, inhalent pores, ostia, and osculum. In a simple sponge cut like the diagram identify the same structures. Do the same for the toilet sponge.

Study a diagram of a portion of the body wall, considerably enlarged. Identify the endoderm and ectoderm cells, the spicules or fibers, and, among the spicules or fibers, irregular amoeboid cells, sometimes called mesoderm cells.

Examine a fragment or section of each kind of sponge under the microscope. Notice the arrangement, shape, and length of the spicules and of the fibers.

Test both kinds of sponges by dropping a bit of each into weak acid, and noting the results. Also burn a bit of each and notice the odor.

Questions.

1. What is the shape of a simple sponge? What enables a mass of cells to retain such a definite shape?

2. What seems to be the composition of the skeletons? Why is one type of skeleton rigid and the other elastic?

3. Since sponges are attached for most of their lives to stationary objects, suggest means for obtaining food and oxygen, and for getting rid of waste matter.

4. Although individual cells are sensitive, a sponge as a whole is not. What connection has this fact with the fact that sponges are stationary?

5. Compare simple and complex sponges.

Suggested drawings.

a. A view of a simple sponge. Label everything shown.

b. A diagram of a simple sponge split in halves. Show by arrows the path followed by the water as it passes through the sponge.

c. A few spicules.

d. A few fibers.

Summary of Important Points in the Study of Sponges

1. What are two functions of the spicules or fibers?

2. What are at least two of the functions of the endoderm cells?

3. What can you suggest as functions for the ectoderm cells?

4. In what cases do cells show "team work" in accomplishing an object?

5. What degree of specialization is indicated by the fact that the cells may exchange positions and functions?

6. What work can any single cell of a sponge do? Compare the work done by such a cell with that done by a paramecium.

7. What work can a whole sponge do? Compare that with the work done by a paramecium.

Review and Library Exercise on Sponges

1. What are the distinguishing characteristics of Porifera?

2. Sponges were once supposed to be plants. In what respect are they plant-like? What made students finally class them as animals?

3. How do sponges reproduce? How are they distributed to new locations?

4. Where, as to depth of water, do most sponges grow? Where, as to oceans? Where, as to latitude?

5. What are some of the difficulties which confront a stationary animal? How are they overcome?

6. To what class of sponges do the "toilet" sponges belong? Why?

7. What conditions are necessary for toilet sponges to thrive? Where are the best ones found? Where are they most numerous? How are they collected? How are they prepared for market?

8. What is man able to do toward raising good sponges for market?

9. Using reference books and museum specimens, describe some especially odd sponges.

3. A STUDY OF COELENTERATES

To show cells working together more definitely than in Sponges

A Study of Hydra

Materials.

Living hydras in permanent aquaria, undisturbed. Living hydras in small aquaria, i.e. tumblers, test tubes, watch glasses, etc., with pieces of water weed and if possible some of the microscopic animals found in water where hydras are abundant. If kept cool, hydras may live several days in such aquaria. Permanent slides of hydras; some whole, some in sections, and some showing the organs of reproduction.

Definitions.

Proximal end, the end by which an animal is attached to an object.

Distal end, the end opposite the proximal end.

Tentacles, slender projections around the distal end.

Mouth, the opening through the distal end, into the central cavity.

Bud, a small hydra or other coelenterate growing out from the wall of the parent.

Mesoglea, a thin, gluey partition, without wandering cells, between the ectoderm and the endoderm.

Nettle cells, very small cells, chiefly in the tentacles, easily identified in permanent preparations as clear cells with small hairs projecting from them. See text-books for details of their structure.

Spermary, the region or organ where the sperm cells are formed.

Ovary, the region or organ where the egg cells are formed.

Coelenterates (hollow bowels), sac-shaped animals, the digestive tract having only one opening; the body wall is of two layers.

Directions.

Take a small aquarium to your table, set it down carefully and leave it undisturbed. Identify a hydra and watch it for some time.

Observations on the living animals.

1. Describe the size and shape of a hydra when expanded. Disturb it slightly by shaking the aquarium a little, and describe its shape when contracted. Notice also the flexibility of the body. What do you infer concerning the hydra's possession of a skeleton? What advantage can it be to have a body so flexible?

2. How many tentacles has the hydra that you are studying? What does the hydra do with these tentacles when it is expanded? What is the probable object of such actions?

3. How does a hydra respond to contact? What seems to be the object of such a response?

4. Notice the location of the hydras in the large, undisturbed aquaria. Where are they placed as regards the light side of the aquarium? Of what value is such a response to light in their case?

5. How can a hydra locate the small animals which are its food?

How can it capture them?

6. What motions may a hydra perform, while remaining attached by its base? What are the results of these movements?

7. If you have happened to see a hydra move from one place to another, describe the process. If not, give the facts which lead you to believe that it is able to do so. Suggest all the methods you think it may be able to use. What is your opinion of the hydra's power of locomotion? Of what use is it in getting food; in escaping enemies; in following the fluctuations of the water supply? If you had to class the hydra as either one, would you call it a stationary or a

locomotory animal?

8. Study budding hydras. Compare the bud with the parent hydra as to size, form and number and size of tentacles. Notice whether the bud moves independently or only with the parent. When does it separate from the parent?

9. In hydras collected late in the fall you may see another method of reproduction. If such material is at hand, notice small swellings near the proximal end and others near the tentacles. Eggs are produced in the lower one, the ovary, and sperm cells in the upper one, the spermary. Refer to your text-book for further details.

Details of structure.

1. Using an entire mounted specimen and a section of hydra, identify the body wall and the central cavity. What is the extent of the central cavity? (Examine both the body and the tentacles.) Where does it open to the outside? What do you think is its use?

2. In the body wall, identify the endodermal and ectodermal layers of cells, separated by the mesoglea, which is usually stained more deeply. Study these cell layers carefully. What work ought each to do? What can you discover in its structure which would fit each layer to do its work?

3. In the tentacles, identify the nettle cells. Where are they? How are they arranged? About how many of them would be discharged if a small animal were to bump into a tentacle?

Summary of Important Points in the Study of Hydra

1. Name the different kinds of cells in a hydra. Which kind differs most from such a cell as the starfish egg? What work does this specialized cell do?

2. How much of a hydra's body may be set in action by touching a tentacle? Contrast this with the sponge. What do you infer concerning the nervous power of these two animals?

3. Look back over your notes and list the different kinds of work a hydra can

do.

4. Can it do any more kinds of work than a paramecium or a sponge can? If so, give further details.

5. Can it do any of its work in any better way? Would you expect it to be able to? Why, or why not?

Suggested drawings.

a. Hydra undisturbed, and hydra after being touched or shaken.

b. A hydra in successive poses to show its flexibility.

c. A hydra taking food.

d. Hydras to show reproduction in one or both ways.

e. A section of hydra, showing details.

Comparative Study of Coelenterates

Materials.

Various coelenterates, such as hydroids, hydro-medusae, jellyfishes, sea anemones, corals, sea fans, etc. Since nearly all the coelenterates except hydras are marine forms, these will usually have to be dead specimens, preserved in formalin or alcohol, or put up as permanent preparations for the microscope.

Definitions.

Colony, as used in this group, a number of individuals descended by budding from an original one, and remaining connected.

Polyp, an individual coelenterate; one of the individuals in a colony.

Observations.

1. How large is an individual specimen in the form you are studying? If the form is colonial, how large is the colony or portion of a colony you are studying? Estimate the number of individuals in it. Is the colony free-swimming or attached? If attached, to what is it usually fastened?

2. Compare the individual you are studying with a hydra, as to size and shape of the body, the location of the mouth, and the size, number, and arrangement of the tentacles.

3. Is there a skeleton? If so, describe it. What appears to be its use? In corals, notice the radiating partitions.

4. Has the specimen any nettle cells? If so, where are they located?

5. Are all the polyps of the colony alike? If not, how many kinds are there? How do they differ?

What is each kind best fitted to do? What is the probable result of this differentiation?

6. What kinds of reproduction, if any, does the specimen you are studying show?

Find out from books what other forms of reproduction are sometimes used by this animal.

Suggested drawings.

a. At least one drawing of each coelenterate you study.

Summary of the Comparative Study of Coelenterates

1. How may polyps in colonial forms differ from polyps which live singly?

2. What variations in methods of reproduction are shown in this group?

3. Which of the polyps you have studied shows the greatest differentiation? In what ways?

4. What characteristic do you find common to all the coelenterates you have studied?

Review and Library Exercise on Coelenterates

1. What are the characteristics which distinguish coelenterates?

2. Give the classes of coelenterates, with the characteristics and an example of each.

3. What enables a hydra to stick to a support by its foot?

4. What are the processes in a hydra by which food is captured, swallowed, and digested?

5. What is the chief fact of interest about Hydra viridis?

6. Why do hydras reproduce all summer by budding and in the late fall by eggs?

7. What change would have developed a hydra and its offspring into a plant-like colony instead of into a group of individuals?

8. Why are ctenophores more easily seen in the night than other coelenterates are?

9. What relations may exist between hydroids and hydro-medusae?

10. What are the advantages of a sedentary life? Of a locomotory one?

11. What is meant by the expression "alternation of generations"? Which animals are likely to develop alternation of generations, sedentary ones or locomotory ones? Why?

12. Give at least two differences between hydro-medusae and true jellyfishes.

13. In the association between a hydractinia colony and a hermit crab, what

advantages are derived by the hydractinia? by the crab? Define symbiosis. Give another illustration of it.

14. How are new coral colonies started? How are large colonies formed?

15. What are the conditions of life under which corals can grow vigorously?

16. Where are corals most abundant?

Note.--Show by coloring the regions on a blank map of the world.

17. How may corals form a reef? Why do they, as a rule, form a reef instead of adding directly to the mainland?

18. Give Darwin's theory regarding the way a coral atoll may have been formed.

19. Where are fossil corals found in abundance? What does their presence prove?

20. What is polymorphism? Give an illustration. What may be a disadvantage of polymorphism? What may be an advantage?

21. In what ways is this group of economic importance?

4. A STUDY OF WORMS

To show cells associated even more closely than in coelenterates, forming tissues and systems of organs.

#A STUDY OF EARTHWORMS#

The Living Earthworm

Materials.

Living earthworms, some of which are left undisturbed from day to day, in damp earth with leaves of various plants scattered upon it.

Definitions.

Anterior end, the head end, usually the leading end.

Posterior end, the end opposite the anterior end.

Ventral surface, the lower surface, usually the one which contains the mouth.

Dorsal surface, the one opposite the ventral surface.

Somites, the rings or segments of which some animal bodies are composed.

Bilateral symmetry, the symmetry usually shown by animals which have differentiated dorsal and ventral surfaces, and right and left sides. Animals which do not have such differentiated surfaces are usually radially symmetrical, but sometimes asymmetrical (without symmetry).

Girdle, the somewhat transparent band frequently found near the anterior end of an earthworm.

Anal opening, the posterior opening of the food canal.

Setae (singular form, seta), small bristles or stiff hairs. In the earthworm these are set in the body wall at definite intervals, and aid in locomotion.

Cuticle, in the earthworm a delicate, shining cover over the body.

Egg capsules, small, light-colored, spindle-shaped sacks, about the size and somewhat the shape of a grain of wheat, containing the eggs or young of earthworms.

Directions.

Take a living earthworm to your table and keep it damp by placing it in a wet tray or upon moist paper. Identify the anterior and posterior ends, the dorsal and ventral surfaces, and the right and left sides. Identify also the somites and the girdle, the mouth with its projecting lip, and the anal opening.

Observations.

1. Watch a living worm for some time. Does it seem to have a definite object in its moving? If so, what is it? Upon what sense or senses does it seem to depend for guidance? Which end usually leads? Why?

2. Over what sort of surface does it move most easily? Why? Watch it closely for some time and discover how it is able to move from place to place. (Suggestion. What is the function of the setae in this process? How can you explain the alternate contraction and expansion of parts?)

3. From time to time, for perhaps a week, examine the leaves which were scattered where the worms could reach them. Have the worms moved them about at all? If so, where are the leaves left? Have any been eaten, in part or entirely? If so, is there any evidence of selection, either as to the kind of leaf or the portion of leaf eaten? If earthworms select food, what senses would be useful for the purpose? Have you any evidence that earthworms possess such senses?

4. Looking through the dorsal wall, notice the meandering red line, seen more easily in some regions than in others. This is the dorsal blood vessel. How long is it? Where is it wider? Where narrower? Notice its pulsations. How many times per minute does it pulsate? In which direction is the blood forced? Is there a corresponding ventral blood vessel? Place a small worm between two pieces of glass, so that you may see through it more easily, and identify the blood vessels encircling the digestive canal, near the anterior end. These are the so-called "hearts" of the earthworm. If possible, decide in which direction the blood flows through them.

5. The food canal, or alimentary canal, lies underneath the dorsal blood vessel, and is usually easily seen, especially if it is full of food. Notice it when the worm is fully stretched and again when it is contracted. How long is the canal? Why does it wrinkle when the worm contracts? Where does it open to the outside? Why does it need to?

6. Where do you infer respiration must take place in this animal? Why do you think so? What fits this surface for such a purpose? Why does an earthworm

seem so uncomfortable when it is too dry?

7. Where do earthworms live? What conditions are necessary in their habitat?

8. When do earthworms usually leave their burrows? Why at that particular time rather than at another? Why does "the early bird catch the worm"?

9. What enemies do earthworms have? How are they protected against these enemies?

10. If you have found egg capsules when collecting worms, describe them.

External Morphology of Earthworms

Materials.

Preserved earthworms, the larger the better.

Observations.

1. In what respects are the dorsal and ventral surfaces alike? In what respects different? Why?

2. Why are the right and left sides alike?

3. In what respects are the two ends alike? In what different? Why?

4. How many somites are there from the anterior end to the girdle? How many under the girdle? How many from the girdle to the posterior end?

5. Where are the setae located in a somite? How are they distributed over the body?

Suggested drawings.

a. An earthworm, dorsal aspect.

b. An earthworm, ventral aspect.

c. An outline diagram of a cross section, to show the location of the setae, the blood vessels and the alimentary canal.

Internal Morphology or Anatomy

Materials.

(1) Preserved earthworms, as large as you can obtain. (2) Cross sections of earthworms. (3) Longitudinal sections of earthworms.

Definitions.

Body cavity, the space between the body wall and the alimentary canal.

Septa (singular, septum), the thin walls between somites, seen when the worm is opened.

Pharynx, the hard-walled, rather bulbous, anterior portion of the alimentary canal.

Esophagus, the portion of the alimentary canal extending back from the pharynx with thinner walls and smaller diameter.

Crop, the short, wide portion of the canal back of the esophagus.

Gizzard, the hard-walled, short region, just back of the crop.

Stomach-intestine, the portion of the canal reaching from the gizzard to the anus.

Ventral nerve cord, a light-colored thread lying against the inner surface of the ventral body wall.

Nerve ganglia (singular, ganglion), slight swellings on the ventral nerve cord.

Nerve ring or collar, a pair of nerves extending from the ventral nerve cord around the pharynx to a pair of ganglia (often called the "brain") in the dorsal region of the anterior end.

Kidney tubes or nephridia, the excretory organs of the earthworm, occurring as slender, paired tubes in nearly every somite.

Directions.

Select a large worm and cut carefully through the body wall along one side, midway between the dorsal and ventral surfaces, from the anterior end to the posterior. Lay the worm on any convenient fairly soft surface (a piece of pine, cork, peat, paraffin), preferably under water, and pin out the walls so that you can see into the interior.

Identify the structures defined above, as well as the dorsal and ventral blood vessels and the "hearts."

The nephridia are not easily distinguished, though they are very numerous. They are long, slender, coiled tubes, two in each somite, lying in the body cavity, one on each side of the alimentary canal. If possible, identify them.

Notice that most of the internal organs are free from the body wall, lying free in the body cavity.

Questions.

1. What is the extent of the body cavity, anteriorly and posteriorly? What is its shape?

2. What, in general, is the shape of the food canal? How many external openings has it?

3. Into what regions is the food canal differentiated? Suggest one advantage of having these specialized regions.

4. How is the alimentary canal of the worm kept away from the body walls? Why have it thus supported?

5. What is a septum? How many septa are there? What vessels and tubes pass

through a septum?

6. Locate the nerve cord. How long is it? How frequently do the ganglia occur on it? Which end of the living worm is the more sensitive. Suggest the connection between this fact and the location of ganglia.

Suggested drawings.

a. Earthworm, showing structures mentioned in this study.

Details of Structure--Microscopic Anatomy

Materials.

Sections of earthworms, preferably both cross sections and dorso-ventral, longitudinal ones.

Directions.

In a section under a simple lens, identify the dorsal and ventral surfaces, the body wall, the body cavity, the alimentary canal, and, if possible, the dorsal and ventral blood vessels and the ventral nerve cord.

Under a microscope identify the same structures. Notice that the body wall consists of three layers of cells: an outer single layer, the epidermis; a middle layer, the circular muscles; and an inner one, the longitudinal muscles.

The nephridia show as loosely scattered fragments in the body cavity, at the right and left of the alimentary canal.

If you happen to have a section which shows one or more setae, identify the muscles which operate it, and the group of glandular cells at its inner end, which are known as setigerous (from seta) cells.

Questions.

1. Describe the epidermal cells. What is their probable function? Among them notice larger cells, clear and rounded. These are the mucous (slime) cells.

What is the use of mucus to the worm?

2. Describe the muscle cells. In which direction do the muscle fibers extend? What is their function? Which layer of muscle cells is thicker, the circular or the longitudinal? Why should it be?

3. Notice the cells in the walls of the alimentary canal. What layers do you find? How are they arranged?

4. If the section you are studying is a cross section from the region back of the gizzard, the alimentary canal will look horseshoe shaped, indented from the dorsal surface. What is the effect of this indentation upon the amount of surface in the alimentary canal?

5. Study the cells of the nerve cord. How do they compare in size and shape with the muscle cells?

Suggested drawings.

a. A diagram of a cross section, showing the relation of the organs.

b. A diagram of a longitudinal section, at least through the body wall, to show the arrangement of muscle fibers.

c. A drawing of a portion of the body wall, to show details.

Summary of Important Points in Study of the Earthworm

1. Compared with a hydra, how many cells has an earthworm?

2. Compared with a hydra, how much are the cells of an earthworm differentiated?

3. How are these differentiated cells usually arranged with respect to one another? What advantage is there in this arrangement?

4. Recall the kinds of work done by paramecium, sponge, hydra, and worm,

and at the same time consider also the efficiency of each. Can earthworms do any more kinds of work than any of the others? Can they do any more work? Can they do any of it better? Give the probable reasons for this?

Comparative Study of Worms

Materials.

As many different kinds of worms as you can get, living or dead.

Directions.

Identify your specimens. Then study as many as your time will allow, using these general questions for each:--

Questions.

1. How large is the specimen and what is its shape?

2. Can you distinguish a head or a head end? If so, by what peculiarities?

3. State whether the body is segmented or not, and, if it is, whether the segments are alike in form and appearance, i.e. whether the segments are uniform.

4. State whether the animal is bilaterally symmetrical, radially symmetrical, or without symmetry.

5. Compare this worm with the earthworm as to sense organs.

6. What organs for respiration has it?

7. What special protective devices has it?

8. If possible, find out and state where this worm lives. What can you see in the structure of this worm which enables it to live where it does?

Summary of the Comparative Study of Worms

1. Name the different worms you have studied. What characteristics have they in common?

2. What different methods of obtaining food do they show?

3. What variations do they show in senses? in sense organs?

4. Which one seems to you best adapted to its habitat? In what ways?

Suggested drawings.

a. One drawing of each worm studied.

Review and Library Work on Worms

1. What are the distinguishing characteristics of worms?

2. Give the classes of worms, and the authority for this classification.

3. What kind of soil do earthworms seem to prefer? Why should they? How do they form their burrows? What are the castings around the mouth of a burrow? How are they placed there?

4. In what ways do earthworms benefit the soil? How great is their effect estimated to be?

5. Give a brief sketch of the life of Charles Darwin, noting especially the work he did with earthworms.

Why is Darwin's work on earthworms noteworthy: because it is such a large proportion of the work he did, or because it is so much of the work which has been done on earthworms?

6. How are earthworms protected against the cold of our winters? What limits the northern range of earthworms?

7. Where are earthworms found geographically? Why are they so widely

distributed? By what means are they extended from one locality to another?

8. How do earthworms reproduce? What care do they take of their young?

9. What tissues or organs of earthworms correspond in function with the ectoderm of hydra; with the endoderm? Why does an earthworm need a system of blood circulation more than a hydra does?

10. Contrast the number of openings in an earthworm's alimentary canal with the number in a hydra's digestive cavity. Which plan seems a better one? In what respects?

11. Contrast a cross section of hydra with one of earthworm as to the number of cavities. Which seems to you the better plan? Why?

12. Why does a nereis need more respiratory surface than an earthworm does?

13. Comparing earthworm and nereis, in what respects is the earthworm degenerate? How does it manage to succeed so well with such a degenerate body?

14. What is a parasite? How many hosts does a typical parasite require for its development? Which host is known as the intermediate one?

15. Trace the history of a tapeworm from the egg to the adult. At what stage are they most likely to be destroyed? What provision is there for this? What advantages are there to the host in the fact that a tapeworm's egg cannot develop in the original host? What advantages to the parasite?

16. What organs has a parasite lost, if it ever had them? How does it succeed without them? What connection is there between parasitism and degeneration? Can you decide which is cause and which is effect? If so, which is?

17. Why do worms so easily become parasitic? What advantages are there in becoming a parasite? What disadvantages?

18. What is radial symmetry? Name two animals which show it. What is bilateral symmetry? Name two animals which show it. What is the relation

between locomotion and symmetry?

19. What is meant in biology by the term "regeneration?" To what extent have we this power? To what extent have hydra and earthworm? What are the results of this power?

20. Name various methods of locomotion among worms. Give examples. Name a fixed or sedentary worm.

21. What is the economic importance of worms? Consider here not only earthworms and tapeworms, but also the stomach worms of sheep, liver flukes, trichinae, hookworms, vinegar eels, and as many others as you have time and books to look up.

5. THE CONNECTION BETWEEN STRUCTURE AND FUNCTION

A Review of the Work done on the First Four Groups of Animals

Review all your studies on the protozoa, sponges, coelenterates, and worms. Write the results in the following summary:--

1. What work, i.e. labor, must an animal do to live?

2. How many cells are necessary to do this work?

3. When this work is divided among a number of cells, what is the effect upon the quantity and quality of work accomplished?

4. When this work is divided among a number of cells, how does the structure of the cells show it? How does the arrangement of the cells also show it? Give examples.

5. The technical expression for this specialization of cells, giving them different functions, is "division of labor." Formulate a clear definition for this expression, giving an example to illustrate it.

6. Is division of labor a good thing for an animal body, or is it not? Give reasons for your opinion, with examples for illustration.

CHAPTER IV

ADAPTATION TO SURROUNDINGS

#A STUDY OF CRUSTACEA#

#To Show the General Adaptation of an Invertebrate to its Surroundings#

1. A STUDY OF CRAYFISHES

Materials.

Crayfishes, living and preserved. Some of the living crayfishes should be established in conditions as natural as possible i.e. in an inch or so of fresh water, with rocks, weeds, etc., and left undisturbed. Small crayfishes are desirable to show locomotion in water.

Living Crayfishes

Directions and observations.

1. Observe living crayfishes in their usual habitat or in a large aquarium, without disturbing them, and see where they stay when they are free to choose. Notice their position. What senses are on guard? What is the color of the head and claws? How may this color aid the animal in getting food or in escaping enemies? Why is the color of the posterior region less important than the color of the anterior?

2. Offer them bits of meat. If one takes food, notice the appendages it uses. How does it discover the food? With what appendages does it grasp the food? How is the food conveyed to the mouth? With what senses, if any, does the animal test the food as it eats it?

3. If the crayfishes are in plenty of water and you startle them in any way, some of them may swim. Watch for such an occurrence and notice it carefully. How is swimming accomplished? Which end leads in swimming? How far does the animal swim at a stroke? How long does it continue to swim? Where

does it go? Does it see where it is going? For what purpose would this method of locomotion be useful?

4. Place a living crayfish in a tray with water to cover it, and take it to your table. Watch the crayfish as it walks about in the water, then take it out and let it walk out of water. Compare the two processes. What causes the differences?

5. How many appendages are used in walking? What order, if any, is there in moving the legs? Which method, walking or swimming, does it use in going to some particular spot, e.g. in going to find food or cover? Why?

6. Gently turn the animal on its back and watch the movements of its appendages as it rights itself. Which appendages does it use and how does it use them? How can it manage to use so many appendages in harmony, for one result?

7. For what different purposes have you seen the crayfishes use their large claws? For which does the claw seem best fitted? Can you think of any change which would make it more efficient for its main purpose? If so, describe the change and tell how it would work.

8. Test the distribution of the sense of feeling. Is it anywhere especially acute? If so, where? Why have two pairs of feelers? Where is each pair carried when the animal is at rest; when it is in motion? How much territory can the two pairs guard?

9. Touch the eyes. Compare their sensitiveness with that of your own eyes. What movements can the eyes perform? How are they protected? What range of territory can they guard?

10. What other senses, if any, do you think a crayfish has? Why do you think so?

11. Early in the spring crayfishes may be found carrying eggs or young. If such a specimen is at hand, notice where and how the eggs or young are attached. How many are there? How are they cared for? Can the young crayfish let go? If removed, can they attach themselves again? How much care does the mother give them when they are removed?

Morphology of a Crayfish

Definitions.

Cephalo-thorax, the anterior half of the body, divided into the head and the thorax.

Cervical groove, the groove dividing the head from the thorax.

Abdomen, the posterior half of the body, consisting of a number of somites.

Note.--The central part of the tail fin is usually included as a somite.

Carapace, the continuous shell-like portion of the exoskeleton covering the cephalo-thorax.

Rostrum, the sharp projection of the carapace at the anterior end.

Gill chamber, a pocket on each side of the thorax, covered by a flap of the carapace.

Appendages, paired structures attached to the body. They are named as follows:--

Eyestalks. (These are not classed as appendages by all students.)

Antennules, the small feelers.

Antennae, the large feelers.

Mandibles, the jaws, one on each side of the mouth.

Maxillae, the two pairs of small mouth parts just back of the mandibles.

Maxillipeds, the three pairs of appendages between the maxillae and the large claws.

Chelipeds, the large claws or pinchers.

Walking legs, the four pairs of appendages back of the chelipeds.

Swimmerets, the appendages on the abdomen.

Openings, five on the ventral surface, as follows:--

The openings from the excretory organs, through small white cones on the bases of the antennae.

The mouth, farther back, between the maxillipeds.

The anal opening, in the last segment of the abdomen.

The opening from the reproductive gland, toward the posterior part of the thorax.

Observations.

1. How large is your specimen? How does it compare in size with other crayfishes in the laboratory?

2. Describe the shape of the body, contrasting the anterior end with the posterior, and the dorsal surface with the ventral.

3. Study the amount of motion permitted in different parts of the body. What prevents motion? What permits it? Where is the body most flexible? Why? Where is it most rigid? Why?

4. How much of the surface is covered with exoskeleton? What arrangement is there to permit the animal to feel contact?

How can the animal grow with such an exoskeleton?

5. Place a dead crayfish in dilute acid for a few hours. What is the result? What has the acid done? Explain the fact that crayfishes are often found alive and well with a soft shell?

6. Compare the cephalo-thorax with the abdomen as to size, shape, and flexibility.

7. How many somites are there in the abdomen? Which way does it bend? Study the somite shells on every side and then state what there is in their construction which determines the direction and amount of their motion. How are the somite shells arranged to protect the body during bending? How is the ventral surface of the abdomen protected?

8. Where are the appendages attached? Study a walking leg and describe its general construction, the number and kind of joints, the direction of motion in each joint, and the range of motion for the whole leg. Study an antenna in the same way. What methods are used in the crayfish to secure a wide range of motion? To secure flexibility?

9. Carefully split a crayfish into right and left halves. To do this, first cut through the ventral exoskeleton from end to end with scissors, then with a sharp knife or razor cut through to the dorsal exoskeleton and cut that with the scissors. Study one half, to get a better idea of the attachment of the appendages. These may then be removed and placed in order on a piece of paper upon which a list of the appendages has been written.

10. How many pairs of appendages are there? How may they be grouped according to location; how grouped according to function? How many pairs are there in each group?

11. What similarities of structure do you find in nearly all of the appendages? Assuming a swimmeret of the third, fourth, or fifth somite to be the least changed from the primitive type, what changes were necessary to make the sixth swimmeret; the third maxilliped; the walking legs; the antennae; the antennules?

12. Remove the part of the carapace which covers a gill chamber. What are the boundaries of the chamber? Where does it open to the water?

13. Describe the appearance and the texture of a gill. How are the gills kept moist when the crayfish is in water; when it is on land? Why should they be

kept moist?

14. Would you class the gills as external structures or as internal? Why do you think so? To what are they attached? How are the gills affected by the motion of the legs?

15. What work goes on in the gills? How is the supply of oxygen renewed? In this connection, try a live crayfish, kept quiet in water just about deep enough to cover it. Float bits of paper near it or carefully place a drop of ink in the water near it. By some such method currents of water may usually be shown, and their direction determined. Consider also the habitual motions of mouth parts and swimmerets, the bubbles sometimes seen when a crayfish is dropped into water and the habit crayfishes have of lying on one side, close to the surface of the water.

Summary of the Study of Crayfishes

To summarize your study, write a connected account of the relations of crayfishes to their environments, under the following topics:--

1. What are the varying conditions in their surroundings which crayfishes must meet? Which are most important?

2. What conditions must be maintained in order that crayfishes may succeed, i.e. may live and reproduce?

3. How does a crayfish know what are the conditions around it?

4. How is it fitted to meet these conditions? (Answer in the following details):-
-

a. How wide a food range has it, i.e. how many kinds of food does it eat? How does it find its food? How does it reach it? How does it take its food? How does it make food small enough to be eaten?

b. What are the organs for taking in oxygen? Where are they? How are they attached? How is the supply of oxygen kept up? How are the organs kept from drying, from clogging, and from mechanical injury?

c. What ranges of temperature can crayfishes endure? What temperature is best? How do they avoid fatal extremes?

d. What are the enemies of crayfishes? What protection against these have they?

e. How often do crayfishes reproduce? About how many times during a normal lifetime? About how many eggs are there and how many of them hatch? What care is given to the eggs and to the young? About how many of the young reach maturity? (Suggestion. Do the crayfishes of a region vary noticeably in numbers from year to year?)

5. What limits the range of crayfishes, north and south? What limits it on land? What in water?

6. When the crayfishes of a given locality are not well adapted to it, what can they do?

Suggested drawings.

a. The whole animal, dorsal surface, preferably without appendages.

b. One of each pair of appendages, except where they duplicate.

c. The tail-fin. Label the sixth swimmerets, the sixth and seventh somites.

d. The gill chamber, with gills in position. Show circulation of water by arrows.

e. A gill, to show construction.

2. COMPARATIVE STUDY OF CRUSTACEA

Materials.

Get together all the different specimens of crustacea you can collect, and identify the material. Then study each specimen as follows:--

Questions.

1. Briefly describe the exoskeleton, if there is one.

2. What region or regions are clearly segmented?

3. How much of the body is covered by a carapace?

4. Has it segmented appendages? Classify the appendages as to their use.

5. Are the cephalo-thorax and abdomen equally developed? If not, which is more developed?

6. How many antennae has it? Are the eyes stalked, or are they sessile?

7. What organs of respiration has it? Where are they attached?

8. How many thoracic appendages has it, if any?

9. What methods of locomotion does it use?

Summary.

1. Does this animal seem to be adapted to life on land or in water, or both? Give your reasons for your opinion.

2. What characteristics are common to all the crustacea you have studied?

3. REVIEW AND LIBRARY WORK ON CRUSTACEA

1. What are arthropods?

2. Give the classes of arthropods with an example of each.

3. What are the distinguishing characteristics of the class crustacea?

4. In what respects are most of the appendages of the crayfish homologous,

i.e. alike in the plan of structure? Which do you consider the simplest, and why do you? Which do you consider the most specialized, and why?

5. Which somite of the crayfish is without appendages? How many somites are there in a crayfish's body, if each somite bears only one pair of appendages, as many scientists believe? How many of these are in the head; thorax; abdomen?

6. Compare the nervous system of the crayfish with that of the earthworm as regards efficiency. Upon what do you base your answer?

7. Name two points in which earthworms and crayfishes are alike. Name three in which they differ.

8. How are crayfishes caught for market? Where do those sold in Chicago usually come from? How are they shipped?

9. Compare the young forms of a crayfish and a crab.

10. Describe any five different crustacea.

11. Describe the work done by the United States government and by the state governments to protect and to perpetuate the lobster. Why is it thought necessary to do this?

12. Discuss the process and the advantages and disadvantages of molting, as seen in the crustacea.

13. Name two advantages in having such a shell as crustacea have. Name two disadvantages. On the whole, is such a shell favorable to an animal's chances of success or is it not?

14. Give the curious myth about goose barnacles.

15. What crustacea are parasitic? Give an account of one.

16. Why are barnacles classed among crustacea? Where were they once classed? Why may they be considered degenerate, even though not parasitic?

How do they manage to succeed? What is their economic importance? How are their effects checked or prevented?

17. Describe some of the odd means of self-protection shown among crustacea.

18. Describe a compound eye. Give two theories as to what can be seen with a compound eye. Why do we not know, instead of theorizing?

19. What is the economic value of the very small crustacea?

20. Discuss the value to man of the various forms of crustacea.

CHAPTER V

ADAPTATIONS FOR PROTECTION FROM ENEMIES

#A. THE EXOSKELETON#

1. THE CLAM--A TYPE OF MOLLUSCA

To Show the Effect of a Heavy Exoskeleton

Materials.

Living clams in aquaria, with enough moist sand to cover the clams, preserved clams, sets of matched clamshells, a few shells with the hinge unbroken, evaporating dishes, hydrochloric acid.

Definitions.

Mollusca, a branch of the animal kingdom including those animals with soft, unsegmented bodies, inclosed in two folds of skin known as the mantle. They are often called shellfish as most of the forms have a shell.

Lamellibranchiata or Pelecypoda, names given to the class of Mollusca to which the clam belongs. The former term refers to the broad, flap-like gills and the latter to the hatchet-like foot.

Valve, one of two parts of the clamshell.

Hinge ligament, the elastic structure which fastens the valves together at the dorsal margin.

Umbones, a pair of elevations near the anterior end of the shell.

Lines of growth, concentric lines around the umbones.

Siphons, two openings at the posterior end of the clam, the upper opening is the excurrent opening and the lower the incurrent. In the salt water clam the siphons form a long tube, usually called the "neck."

Hinge teeth, projections near the dorsal margin on the inner surface of the shell. The anterior irregular structure is the cardinal and the more posterior blade-like structure is the lateral tooth.

Muscle scars, scar-like markings on the inner surface of the shell indicating the point where muscles were attached. The large scar just in front of the cardinal tooth is the anterior adductor muscle scar, and the one just back of the lateral tooth is the posterior adductor muscle scar.

Pallial line, a line connecting the two muscle scars.

Mantle, folds of skin covering the body of the clam and lying close to the inner surface of the valves.

Foot, a hatchet-shaped structure extending from the ventral edge of the body.

Gills, broad flap-like structures for respiration, situated each side of the body in the mantle cavity. They consist of a double fold of membrane through which run many perforations lined with cilia. The waving of these cilia cause the current of water needed for respiration.

Palps, small flap-like structures near the anterior end of the clam. They surround the mouth. On their surface are cilia which cause currents of water toward the mouth.

Adductor muscles, large muscles extending from valve to valve.

Observations.

Identify anterior and posterior ends, dorsal and ventral surfaces, right and left sides.

1. Why may a clam be called a bivalve?

2. What is the position of the clam in the mud? What is the position of the foot if the clam is undisturbed? Are the two valves tightly closed or slightly open at this time?

3. What changes take place in the shell as the clam grows? What markings on the surface of the shell indicate this?

4. Where is the clam sensitive to touch or tactile stimulus? Why has the clam no eyes? Zoologists have found a structure in clams which they have supposed to be an ear. Where do you think the structure is located? Why is the clam successful without eyes? (There are many bivalves which have them.)

5. Examine several clams until you find some with enlargements in the gills. Break off a small part of an enlargement with your forceps and examine under the compound microscope. Describe what you see.

6. Drop some powdered chalk or carmine in the water just above the siphon, watch the siphons for several minutes, and note what happens. What do you conclude to be the use of the siphons? Recalling what took place in sponges, what would you suggest as the probable cause of these currents? What does the clam thus probably obtain? How do the two siphons differ? Why?

7. Place a clam in water sufficient to cover it and heat slowly to about 40 degrees Centigrade, until the valves open slightly. Remove and proceed as follows: Raise one valve, separate the mantle from it, and then cut through the two large firm structures (adductor muscles) found at each end. What does the valve do when the muscles are cut? What is the cause of this? State your theory as to how a clam opens and closes its shell.

8. Note the texture of the mantle. How many lobes has it? What is their extent? How are the lobes related to the valves?

9. Remove or lift up one mantle lobe. Identify the soft body, the foot, the gills, the palps, and the mouth. Which of these structures are arranged in pairs?

10. Determine the structure and composition of the shell as follows:--

a. Break a thick clamshell and examine the broken edge. Identify the inner or pearly layer and the outer or chalky layers. What gives color to the shell in the living clam?

b. Burn a small piece of shell in an evaporating dish over a bunsen burner. What is the appearance of the shell after burning? What has been burned, animal or mineral matter? What then is the residue?

c. Place a small piece of shell in acid. What results? Is there a large amount of residue? What constitutes the greater part of the shell, animal or mineral matter?

d. (Optional) Devise some method and determine the approximate per cent of mineral and of animal matter in the clamshell.

Summary.

1. Why did we study the clam? (See title of exercise.)

2. How has the heavy shell of the clam affected:--

a. The character of the clam's body,

b. the locomotion,

c. the development of sense organs.

3. What special problems has the clam as regards getting food and oxygen? How are these problems solved?

4. How does the clam protect the young clam during development?

Suggested drawings.

a. Dorsal margin of the clam.

b. Side view of the clam.

c. The clam with one valve removed or lifted back.

d. The clam with one valve and one mantle lobe removed.

e. The edge of a broken shell.

f. Diagram of cross sections.

2. THE SNAIL--A TYPE OF MOLLUSCA

To show Another Type of Exoskeleton

Materials.

Specimens of pond snails, edible snails, and "slugs," and other land snails, and a collection of shells of various types.

Definitions.

Gasteropoda, the name of the class to which the snail belongs.

Spire, the coiled portion of the snail shell.

Aperture, the opening of the shell.

Lip, the edge of the shell forming the margin of the aperture.

Whorl, a single coil of the spire.

Suture, the depression between the whorls.

Foot, the flat disk-like structure on which a snail creeps.

Breathing pore, an opening in the mantle used in respiration.

Lingual ribbon, the rasp or file like tongue of the snail.

Observations.

1. Why is a snail called a univalve?

2. Identify the head and mouth of the snail. Watch the snail feeding and examine the mouth of the snail with a lens. What do you notice? If your aquarium in which the pond snail is living has a green coating (algae) on the side, describe its appearance after the snail has been crawling up and down over it. Explain.

3. How many tentacles has a pond snail? a land snail? Where are the eyes located in each case? What movements of the tentacles do you notice? What is their purpose?

4. How does the rate of locomotion of the snail compare with that of the clam? Find out if the snail can creep backwards or on the surface of the water. Does there seem to be any tendency for the snail to go up and down the sides of the aquaria vertically rather than to the right or left?

5. What does a snail do when disturbed? What is gained by this action?

6. Search for pond snail's eggs on the side of the aquaria. Lift up the bits of cabbage on which the slugs are feeding and search for eggs. Describe what you find in each case, noting the size, appearance, and whether the eggs are laid singly or in masses.

7. Find the breathing pore. Describe its position and appearance.

8. Contrast the various types of shells, and note with care in what respects they differ. Holding the shell with the aperture toward you and the spire

pointing up, determine whether each shell has the aperture on the right (right-handed shell) or on the left (left-handed shell). Is the right-handed or the left-handed shell more common?

9. (Optional) By means of some book in the laboratory, determine the scientific name of each of the snails found in the various aquaria in the laboratory.

Suggested drawings.

a. Drawings to show the pond snail in various positions in the aquarium.

b. A drawing of the slug.

c. At least three different types of snail shell.

Summary.

1. In what respects does a snail show resemblance to a clam?

2. What are the chief points of difference?

3. What reasons can you suggest for the better development of the sense organs?

4. What advantage has a snail over a clam in the matter of getting food?

5. How does the shell of the snail compare with that of the clam as an organ for protection?

3. THE SQUID--A TYPE OF MOLLUSCA

To show the Effect of a Much Reduced or Rudimentary Skeleton

Materials.

Small squids, and a few large specimens for comparison and dissection.

Definitions.

Cephalopoda, the name of the class to which the squid belongs.

Caudal fin, a horizontal structure at the posterior end of the squid.

Chromatophores, irregular cells in the mantle which give color to the squid.

Exhalent siphon, a funnel or tube opening on the ventral side just below the base of the arms or tentacles.

Pen, a remnant of an exoskeleton imbedded in the mantle along the dorsal side.

Ink sac, a sac containing a dark, sticky liquid which may be thrown out through the funnel into the water. The opening is near the inner opening of the funnel.

Observations.

1. What is the shape of the squid? To what is this shape adapted?

2. Identify the head and the well-developed eyes.

3. How many arms or tentacles are there? How are they arranged with reference to the mouth? What do you find on the distal ends of the arms? How do the arms vary as to size? What does the position and arrangement of the arms suggest as to their function?

4. Identify the exhalent siphon. Where may water enter the mantle cavity? Recalling the action of the siphons in the clam, suggest a method by which a squid is propelled through the water. In what direction must it swim?

5. Split the mantle along the ventral surface and spread apart. Identify the long plume-like gills, the ink sac, and the inner opening of the exhalent siphon. How many gills do you find?

Suggested drawings.

a. The squid side view.

b. The squid from the ventral side with the mantle split open, arrows to show direction of water.

Summary.

1. In what ways does a squid show relationship to the clam and the snail?

2. What has a squid gained through the reduction of its exoskeleton? What has it lost? What changes were necessary in its structure to offset the loss of an exoskeleton?

4. A COMPARATIVE STUDY OF MOLLUSCA

Materials.

Specimens of as many different kinds of mollusks as possible, charts, books.

Observations.

1. What is the symmetry?[3]

[3] This and the following questions are to be answered for each specimen. The answers may be tabulated if desired.

2. Is the body segmented or unsegmented?

3. Are lateral appendages present or wanting?

4. Is an exoskeleton present or wanting? If present, is it univalve or bivalve; if absent, what other means of protection has been developed to take its place?

5. Is the animal fixed, or is it free to move? If fixed, in what way? If it moves, what is the method and organ of locomotion?

6. What are the organs of respiration? What is their character?

7. How is food obtained?

8. What senses are probably present? What sense organs are present?

9. What is the habitat?

10. In what ways if any does the animal show degeneration?

Summary.

1. What characters are common to all mollusks?

2. What is the principal means of protection among mollusks?

3. Name three causes of degeneration among mollusks.

5. MOLLUSCA: REVIEW AND LIBRARY EXERCISE

Characteristics.

1. What are the general characteristics of mollusks?

2. Name the principal classes and give the characteristics of each.

Morphology.

3. What is peculiar about the structure of a clam's heart? What is its position? Contrast with the heart of a crayfish.

4. Make cross-sectional diagrams to show the arrangement of parts in a clam: (a) in the region of the umbone; (b) in the region just in front of the posterior muscle; (c) in the region of the anterior muscle.

5. Describe the various types of eyes found in mollusks, and their location.

6. Describe the tongue or lingual ribbon of the snail, and its use.

7. What is the operculum of snails? its use?

Physiology.

8. Describe the circulation of water through the siphons and mantle cavity of a clam. How is it caused? What three uses has it?

9. What are the principal facts about the development of fresh-water clams?

10. Describe the circulation of blood in a clam.

11. What various methods of locomotion are found among mollusca?

Economics.

Write a short account of the following:--

12. Oyster culture.

13. Typhoid-fever and oysters.

14. Clams, scallops, and other edible shellfish.

15. Pearls and pearl fisheries.

16. Fresh-water clams and the button industry.

17. Sepia, Tyrian dye, etc.

18. Harmful and useful mollusks.

19. The work of U. S. Fish Commission in propagating clams.

Natural history.

20. Give the class, habitat, and some important fact about each of the following: Pectens; wing shells; Tridacna gigas; abalones; limpets; oyster drill; periwinkle; mussel; cuttle fish; octopus; nautilus; argonaut.

6. A COMPARATIVE STUDY OF EXOSKELETONS

Materials.

Charts, specimens, etc. Since this is partly a review exercise, your notes and drawings of invertebrates should be at hand.

Definitions.

Exoskeleton, a protective covering developed on the outside of an animal.

Questions.

1. What are foraminifera; radiolaria? How do they differ from other protozoans? Of what two substances are the shells of protozoans composed?

2. How are the spicules formed in a simple sponge? What are glass sponges? Give reasons why the skeletons of sponges may or may not be considered exoskeletons?

3. What are stone corals? What is the relation of the coral polyp to the skeleton? What is the appearance of the coral when expanded as compared with its appearance when contracted? Of what substance is the coral composed?

4. Describe the exoskeleton of a starfish. Contrast the exoskeleton of the sea urchin and the starfish. Why does a sea cucumber need no well-developed exoskeleton?

5. What structure in an earthworm may be considered an exoskeleton? What other types of exoskeletons are found in segmented worms?

6. Of what substance is the exoskeleton of arthropods composed? What additional substance is found deposited in the shell in the case of crustaceans? What advantage in the arthropod type of exoskeleton?

7. Why are mollusks so commonly called "shellfish"? What advantage in the mollusk type of skeleton? What disadvantages?

Summary.

1. What type of exoskeleton is common among invertebrates?

2. What are the general purposes of exoskeletons?

3. What is the explanation of the various forms of exoskeletons found?

4. Of what substances are exoskeletons composed?

#B. PROTECTIVE COLORATION#

#To show how Color may be Protective#

Materials.

Specimens such as the Kny-Scheerer mimicry collections, diagrams, etc.

Definitions.

General protective resemblance, the general resemblance between the color of an animal and its surroundings.

Variable protective resemblance, the changing of the color of an animal to correspond to the change in its background.

Special protective resemblance, the resemblance of an animal to some object found in its background in color and form.

Mimicry, the resemblance of an unprotected animal to a well-protected one.

Warning colors, bright colors which protect animals by causing other animals to avoid it.

Questions.

1. Show how the transparent color of a paramecium, the green color of a

cabbage worm, or the green color of a certain species of hydra may result in protecting an animal from its enemies. Mention as many other examples as you can.

2. What is gained by the ability of a squid to change its color? How is this change brought about?

3. Explain the protective coloration of the following: Dead-leaf butterfly, walking stick, geometrid larva. Hunt up other examples.

4. Explain the protective coloration in the following: Hover flies, clear-winged moths, viceroy butterflies.

5. Make a list of several invertebrates that are protected by their bright color. Explain the reason for the bright color.

6. How may the difference between the color of the upper and lower surfaces of animals be explained on the basis of use to the animal?

7. (Optional) Find out some other uses of color to an animal aside from protection.

Summary.

1. Name four uses of color.

2. Name four ways an animal is protected by being like its background.

3. Name one way it is protected by being unlike its background.

4. What disadvantages in this method of protection?

#C. ANIMAL ASSOCIATIONS#

#To show Another Method of Protection from Enemies#

Materials.

Specimens, charts, etc., illustrating animal associations.

Definitions.

Animal communities, associations of many animals of the same species in communities in which there is a greater or less division of labor.

Gregarious, associations where there is but little division of labor.

Parasitism, an association where one animal lives at the expense of the other. The animal on which the parasite lives is called the host. If there are two hosts during the life cycle of the parasite, the second host is called an intermediate host.

Symbiosis, an association where two animals live together in mutually helpful relations.

Commensalism, an association where two animals live together in relations not mutually helpful but without injury to either.

Observations and questions.

Note.--To find answers to many of these questions it will be necessary to refer to the reference books in the laboratory.

1. Examine a specimen of Volvox. Why may this be considered a colonial protozoan and not a many-celled animal? What is gained by the colonial habit?

2. Is the colonial habit common or rare in sponges and coelenterates? What is chiefly gained?

3. Describe the community life in one of the insects in each of the following groups:--

a. ant, honeybee, termite.

b. bumblebee, paper wasp, hornet.

c. mining bee.

d. carpenter bee, mud wasp, digger wasp.

4. Name the host or hosts in the following cases: trichina, liver fluke, malarial parasite, tapeworm, hook worm. Give the life history of one or more of the parasites just enumerated. What is the effect of parasitism on the structure of the parasite?

5. What is the relation between ants and plant lice? Show how this relation is mutually helpful. Mention other cases of symbiosis that you have come across.

6. With what animal are barnacles often associated? What is the habit of the pea or oyster crab? What are "guest bees"? What structure is lacking that is found in other bees? What are often found in the cavities of sponges? Why are these associations called commensalism rather than symbiosis?

Summary.

1. Into what groups can animal associations be divided based upon the number of species concerned?

2. From the standpoint of protection, is this a good or a bad method of protection?

3. What disadvantages can you see in this method of protection.

#D. PROTECTIVE HABITS AND POWERS#

Materials.

Specimens, charts, and books, showing habits of invertebrates.

Definitions.

Regeneration, the power to grow new parts of the body when parts have been lost or injured.

Masking, the covering of an animal by some object or organism so as to hide its identity.

Nocturnal habits, the habit of hiding in the daytime and coming out at night to feed.

Terrifying attitudes, the protective attitudes assumed at times by animals in order to ward off attack.

Observations and questions.

1. How are Sabella and Serpula protected? What advantages and disadvantages in this habit? What changes in structure are associated with this tube-dwelling habit?

2. What two protective habits has the earthworm? Name some other animals that have similar habits.

3. Describe the protective habits of the caddis-fly larva; of the leaf-roller moth. What benefit to the hermit crab is the colony of hydractinia growing on the snail shell which it inhabits? Give other similar cases.

4. Name as many cases of regeneration as you can.

5. What peculiar habits has a puss-moth larva? a dragon fly? Give other examples.

Summary.

1. Name the various protective habits.

2. State any advantages or disadvantages you can with reference to these protective habits.

#E. DEFENSIVE STRUCTURES#

#Another Method of Protection from Enemies#

Materials.

Specimens, charts, books, etc., to illustrate the various defensive organs found among invertebrates.

Observations and questions.

1. Describe the stinging hairs of the paramecium.

2. Describe the action and structure of nettle cells. Where are they located in the case of hydra; of jellyfish?

3. What defensive organs are found among the arthropods?

4. What are stinkbugs? What peculiar organs of defense have the caterpillars of the swallowtail butterflies?

5. Where is the sting of a hornet located? To what in a grasshopper does it correspond? Why does a hornet or bee inflict so painful a wound?

6. What peculiar organ of defense has a squid?

7. Find other examples of defensive structures.

Summary.

1. What advantages have organs of defense as a method of protection?

2. What disadvantages?

#F. THESIS#

#To sum up the Important Points in the Study of Adaptations for Protection#

Directions.

Write a connected account of what you have found out about protection of animals from their enemies, using the following outline:--

1. The struggle for existence--

a. its cause, b. its threefold nature, c. the various kinds of adaptations.

2. The various methods of protection from enemies.

a. The exoskeleton. b. Protective coloration. c. Animal associations. d. Protective habits. e. Defensive structures.

CHAPTER VI

VERTEBRATES

#A. STUDIES OF FISHES#

THE LIVING FISH

Vertebrates adapted to Water Life

Materials.

Living goldfishes or other fishes in small aquaria for individual study and a few fishes in a large aquarium where they have considerable freedom of motion.

Definitions.

Trunk, the portion of the body between the head and the tail.

Compressed, a term used to describe the shape of the body when it is narrower from side to side than from dorsal to ventral surface. When the opposite is true, the body is said to be flattened.

Median fins, the unpaired fins situated on the median line, dorsal and ventral, including the tail or caudal fin, the dorsal fin, and the anal fin.

Paired fins, fins occurring in pairs of which the more anterior are the pectoral fins and the posterior are the pelvic fins.

Fin rays, the framework or skeleton of the fins over which membrane is stretched to form the fins. Fin rays are of two kinds: those composed of bone and those composed of cartilage.

Lateral line, a sense organ extending along each side of the fish in a line indicated by tubes or perforations in the scales.

Gills, respiratory organs adapted for taking oxygen from the water.

Operculum, the flaps covering the gills on each side of the head.

Pigment, a substance which gives color to an object.

Observations.

Locomotion.

1. Watch the fishes in the large aquarium and determine which fins are most used and how they are used (a) in swimming forward, (b) in swimming upward and downward, (c) in maintaining balance, (d) in remaining at rest, and (e) in guiding the movements of the fish.

2. What advantages are there to the fish (a) in the power to open and close the dorsal and anal fins, (b) in having no neck, and (c) in having a compressed form?

3. Enumerate the various ways by which the body of the fish is adapted to rapid movement through the water.

Feeding.

4. What is the food of the fishes you are studying? Feed them and watch them eat. Why is the upper jaw often called a "lip"? What is the shape and size of the mouth when opened in feeding? Does the fish chew its food? Describe in detail the fishes' method of feeding.

Respiration.

5. Identify the opercula and the gill openings. Watch the movements of the opercula and mouth, and determine what movements are concerned in breathing and their order. Describe in detail the circulation of water used in breathing and how it is caused.

Sense Organs.

6. Identify the eyes, nostrils, and lateral line. How many nostrils are there and where located? What is the position and extent of the lateral line?

7. Describe the location of the eyes. What is the shape of the outer surface of the eyes? Why this shape? Can the eyes be moved, i.e.can they be rotated, rolled, or retracted? From what direction might an enemy approach without being seen? How would such an enemy be detected?

Protection.

8. With what protective structures is the body covered? Do they hinder the movements of the fish? What are the advantages of the scale covering of fishes over the shell covering of grasshoppers or crayfishes?

9. In what other ways are the fishes you are studying protected against enemies? Since you cannot account for the red color of goldfishes on the basis of use to the fish, then how do you account for this bright color?

The Body.

10. What is the symmetry of the fish? Into what regions is the body divided?

Summary of the study of the living fish.

Enumerate in one column the different adaptations which fit the fish for life in water and in a second column state the special purpose of each adaption.

The External Structure of the Fish

Materials.

Freshly killed or preserved fish in dishes or shallow pans with enough water to prevent drying. Simple or compound microscopes, forceps, and a bristle.

Directions.

Examine the fins and identify the membrane and the supporting rods, or rays, of bone or cartilage. Notice how the ends of the cartilaginous rays keep the membrane from tearing.

Investigate the scales as to their arrangement, number, and size. Remove a small patch of scales along the lateral line to find how they are attached, where the fish's color is situated, and how access to the sensory organs of the line is permitted. Examine a scale under the microscope.

Observe the eyes and identify the parts similar to those of the human eye: lid, lash, tear-duct, cornea, iris, and pupil.

In front of and between the eyes, find the nostrils. By means of a bristle determine whether these are connected and whether they do or do not open into the mouth or the throat.

Questions.

1. Make a list of the fins, classifying them according to their structure.

2. Bearing in mind the differences in structure and consequent action,--what can you say regarding the adaptation of the several fins for protection? for rigidity or flexibility in locomotion?

3. State how much of the body is covered with scales, and where the largest and the smallest ones are found.

4. How are the scales arranged with reference to each other? What benefit is derived from this in protection? in locomotion? If you have noticed any mucus or slime upon the body, state its use.

5. Do the scales or the skin bear the pigment? Give the color pattern of the kind of fish used in class. How would this be useful to the fish in its natural home?

6. Describe the structure of a scale and state how it is attached to the skin. In what way is the lateral-line scale specialized?

7. State how, when the fish is swimming, the nostrils catch odors. By means of a diagram, with arrows show the probable direction of the water current through the nose.

8. State which of the structures of your eye are present in the fish's eye, and which are missing. Could a fish weep? wink? How would a fish sleep?

9. Inasmuch as light penetrates water but a little way, so that objects can be distinguished only within about thirty feet, would the fish be nearsighted or farsighted?

Suggested drawings.

a. A side view of the entire fish, fully labeled.

b. A bony rayed and a cartilaginous rayed fin.

c. A scale, showing its minute structure.

d. A dorsal or a lateral view of the head, showing the sense organs.

The Mouth and the Gills of the Fish

Materials.

The same materials as those used in the preceding exercise may be used here.

Directions.

The mouth, its structure and its action, can be seen by pulling the upper jaw upward and forward until the mouth and the gill chambers open fully.

Examine the structure and action of the jaws, the tongue, the throat, and the teeth on each jaw and on the roof of the mouth.

Investigate the breathing apparatus from the throat side and from the exterior, noting the number, form, and structure of the gills, their attachment and their protection.

The mouth may be kept open by a short splinter or a ball of paper.

The pupil should identify the following structures:--

1. Gill, an organ for breathing the air dissolved in water.

2. Gill arch, an arch of bone or cartilage supporting the gills.

3. Gill filaments, fringe-like structures attached to the gill arches, forming the gills.

4. Gill raker, lateral projections from the gill arches.

5. Gill-slits, openings between the gill arches for the passage of water.

6. Operculum, the flap-like covering of the gills on each side of the head.

Questions.

1. Compared with the size of its body, how wide can the fish open its mouth? What do you infer as to the size of its "bite"?

2. Are the jaws rigidly affixed to the skull? Why should they be so attached, or why not?

3. Of how many pieces is the upper jaw composed? the under jaw?

4. Where are the teeth? Judging from their form, size, and situation, what do you think must be their use?

5. Do you think the tongue is used to assist in mastication? in tasting? in

speech? in swallowing?

6. How many gills are there, and where are they situated? How are they attached? Which one is not free from the body throughout its length?

7. What probably causes the color of the gill filaments? What is there in their number and texture which fits them for their function?

8. What is the direction of the water current through the gill chamber? Of what use are the gill rakers?

9. How are the gills protected?

Summary.

Write a complete account of how the fish eats and how it breathes.

Suggested drawings.

a. A front view of the fish's face, with the mouth fully open.

b. A side view, as above.

c. A ventral view of the head, with both gill-chambers wide open and the gills separated from each other. Indicate currents by arrows.

d. A single gill.

The Alimentary Canal and the Circulatory System of the Fish

Materials.

Small fresh fish, shallow pans or dishes of water, forceps, and scissors.

Directions.

If the instructor has not opened the fish previously, this is to be done by the student as follows: On the ventral side, insert the scissors in the vent (in front of the anal fin) and cut straight forward to a point between the opercula. Care must be exercised in opening the chamber about the heart; this lies between the gill chambers.

The various organs, so far as possible, should be carefully drawn out and separated, in order that their structure may be distinguished.

The pupil should identify the following parts:--

1. Body cavity, the entire internal space, divided by a membrane, false diaphragm, into a large abdominal cavity and a small chamber, pericardial chamber, between the gill chambers.

2. Liver, a large red or pink mass lying at the front end of the abdominal cavity, and divided into two unequal lobes. The gall-bladder, thin-walled and green, may be seen between these lobes.

3. Alimentary canal.

a. Mouth.

b. Esophagus, in the fish a very short tube.

c. Stomach, white and muscular, beginning with a very short esophagus and ending as a blind sac. If it is much distended, open it to see what the fish may have eaten.

d. Small intestine, thin-walled, tubular, and somewhat coiled.

e. Large intestine, a short, thin-walled expansion at the posterior end of the small intestine; usually less than half an inch long.

f. Coeca, from two to several small pouches attached where the small intestine leaves the stomach.

4. Spleen, a reddish brown globule between the folds of the intestine.

5. Swim bladder, an elongated chamber lying against the backbone,

partitioned off from the cavity below by a delicate membrane.

6. Peritoneum, the delicate, silvery membrane which lines the abdominal cavity and enfolds the viscera. Note its spots of pigment.

7. Pericardial chamber, the chamber around the heart; see Sec. 1 above.

8. Heart. As the fish is placed belly upward in the pan the ventricle faces you, pink, conical, and muscular. Posterior to it, on the dorsal side, is the auricle, a membranous sac.

9. Ventral aorta, arising on the anterior surface of the ventricle as a white muscular "cord" (really a tube) which is enlarged close to the heart into a bulb, the arterial bulb. You should follow up this aorta until you see it divide right and left to send its branches outward into the gills, the branches being called gill arteries.

Questions.

1. The fish frequently swallows its food alive. Why should the stomach be muscular? Why is it better that the intestine does not leave the stomach at the end opposite the esophagus?

2. Of what use can the coeca be? What structure of the human intestine do you recall that is at all like them in form or use?

3. How many times the length of the body is the length of the alimentary canal? Does this indicate that the fish is compelled to eat a great deal of poor food or that its food is highly nutritious, so that little need be taken?

4. Near which end of the fish's body is the heart? Is this the usual or the unusual condition among animals you know about? What advantages can you think of in this arrangement?

5. What advantages are there in having the heart in a chamber separated from the other vital organs?

6. Of how many chambers does the heart consist? Why should at least one of

them be muscular?

7. How many times does the blood pass through the heart in making a complete circuit of the body? Would you call this a single or a double circulation?

8. Does the heart force the blood onward or does it draw blood into itself, i.e. is the heart a force pump or is it a suction pump?

9. How is circulation made complete? If the heart is a force pump, is its power sufficient to drive blood through artery, capillary, vein, and into auricle, if the capillaries can stand the pressure, or is another action concerned? If it is a suction pump, why does the blood leave the heart?

Suggested drawings.

a. The body cavity, with viscera undisturbed.

b. The alimentary canal extended.

c. The anterior end of the fish with the sinus held open, to show the general situation of the parts.

d. The heart in its chamber, with the outgoing vessels as far as dissected. Use arrows to show direction of circulation.

e. A copy of some good diagram or chart which illustrates the heart of the fish with the connecting veins and arteries.

Fishes: A General Review and Library Exercise

1. Food and the feeding habits of young and of adult fishes.

2. The diet and habits of cod; lantern-fish; swordfish; ramora; hagfish; angler; gar-pike; sturgeon; shark; sawfish; paddle-fish.

3. The variations, real or apparent, in the breathing habits of the porcupine-fish; the climbing-fish; the lung-fish.

4. Peculiarities in swimming as seen in the flying-fish; the flounder; the sea-horse.

5. Intensity of sound under water, and the corresponding structure of the fish's ear.

6. Light and sight under water (as in 5).

7. Protection of fishes: sting-ray; torpedo; coral-fish; sturgeon; lava-fish; swordfish; sawfish; pipefish.

8. The social instinct of fishes, and "schools."

9. The breeding habits of salmon; eel; stickle-back; sturgeon; whitefish; shark; sea-horse; sunfish.

10. The fishing industries of the Great Lakes or of the cold oceans, with a list of the fishes caught and their values.

11. Fish nets and traps: seine; gill-net; pound-net; trawl, French or English; fish-wheel; fish-weir; spear; dip-net; set-line; spoon; fly.

12. The U.S. Bureau of Fisheries: its locations, its problems, and its methods.

13. The State Fish Commission, as above.

14. Game and fish laws; their purpose and their enforcement.

15. Game fish of the fresh waters; trout, bass, pickerel, and muskellunge.

16. Game fish of the ocean: tarpon, tuna, sea-bass, swordfish, and bluefish.

17. Fish as food.

18. Fish diet and leprosy.

19. Fish diet and parasitic worms.

20. Fish nuisances: carp, catfish, and dogfish.

21. Commercial products of fishes, their preparation and their uses: caviar, shagreen, cod liver oil, isinglass, and glue.

22. The geographic distribution of fishes, with means of dispersal and restriction.

23. The faunal regions of the lake (or ocean), with characteristic forms.

24. Fishes of ancient times; of the Devonian period.

25. The story of the early life of Louis Agassiz; of D. S. Jordan; of C. H. Eigenmann; of Bashford Dean.

26. Goldfish: their origin; how to care for them.

27. Fashions in fish tails, old and new.

28. Development and variation in scales; fashions in scales.

29. The common orders of fishes, with examples.

Primitive Chordates

Materials.

An acorn-tongued worm, a lancelet, a lamprey, a shark, and a perch. If individual specimens are not available, the pupil's text-book and charts are to be used.

Observations.

Acorn-tongued worm: Notice the very simple form and structure of the symmetrical body, the "proboscis," the collar surrounding the neck with its simple rod of cartilage, the marks of internal gills and gill slits extending some distance along the body, and the presence or absence of sense organs. The acorn-tongued worm (Balanoglossus) lives in the sand of the seashore and in shallow water in temperate and tropical regions.

Lancelet: Observe the form of the body, of the fin, and of the mouth; note the presence or absence of sense organs, and find out the number of gills or gill slits. The lancelet (Amphioxus) is similar in habit to the acorn-tongued worm. By day it lies buried with only the mouth exposed, but at night it swims actively about. It is somewhat more confined to the tropics.

Lamprey: Observe here also the primitive or unspecialized form of the body, of the fin, of the jawless mouth, the number of gill slits, and the sense organs.

Shark: Examine the body, noting its form and differentiation into regions, its covering, its fins, mouth, gill slits, and sense organs.

Perch: If you have not already studied the bony fish, the points suggested for the shark will be sufficient for this exercise.

In each case, find out the condition of the skeleton.

Questions.

1. Which of these animals seem most simple in form, and which most complex? Give a reason for your answer.

2. Give the stages which show how the fold of skin develops into separate fins.

3. How does the number of gills and gill slits change in the series? (Give definite numbers.) How may the reduction in the number of gills be compensated for in the amount of surface exposed for the exchange of gases in breathing?

4. How is protection afforded the delicate structure of the gills in the final form?

5. Give the stages in the formation of a definite, symmetrical mouth with jaws of equal size.

6. The presence of sense organs may be taken to indicate that there is an organ of control, or brain. How is the development of this organ like or unlike that of the other structures in the series?

7. For the developing brain and nervous system what protection and support is afforded in each case?

The foregoing questions may be answered in tabular form by arranging the names of the animals in a line and the questions in a column.

Suggested drawings.

a. Acorn-tongued worm, x 1.

b. Lancelet, x 1.

c. Lamprey, x 1/2.

d. Shark: 1, head as far as the pectoral fins; 2, the tail.

e. Perch, as directed for shark.

#B. STUDIES OF AMPHIBIA#

#Progress from Water-living Animals to Land-living Animals#

The Living Frog or Toad

To show how an Animal may be adapted to both Land and Water Life

Materials.

Living frogs in small cages or aquaria; living toads; some pungent liquid, as ammonia.

Observations.

The Body.

1. Contrast the body of the frog with that of the fish as to regions, shape, and compactness. How do you account for the differences?

Locomotion.

2. What kinds of locomotion can a frog use? Why is it difficult for a frog to crawl or walk? How far can a frog jump? How are swimming and jumping accomplished? What do you think is the use of the "hump" on the back?

3. Identify in the legs the thigh, shank, ankle, foot, toes, and web, and in the arm, the upper arm, forearm, wrist, hand, and fingers. State in detail the differences in structure and in position between the fore and hind limbs. How do you account for these differences?

Feeding.

4. Induce a frog or toad to eat by dangling food, such as a piece of raw meat or meal worms, small earthworms, etc., before it. How does it seize the food? What will it eat? How is the mouth adapted to this manner of feeding?

Respiration.

5. The frog has no diaphragm, and therefore no chest cavity; watch very carefully the movements of the mouth, the nostrils, the throat, and the sides of the body to determine how the problem of breathing (how the air is gotten into and out of the lungs) without a diaphragm is solved. Write a detailed account of the frog's method of breathing which shall explain just how the air is forced into the lungs.

6. What would be the effect of propping open the mouth of the frog? Why? Does the frog breathe in the usual manner while under the water? If not, how do you explain its ability to remain under water for a long period of time?

Sense Organs.

7. Investigate the efficiency of the five special senses in the frog by devising

experiments to test each sense; as, for example, giving a frog its liberty on the floor and trying to catch it again, to test the sense of sight. Write an account of your experiments and their results. Which of the senses is best developed? Give reasons for your answer.

8. Compare the eye of the frog with respect to its shape, movements, parts, and protective structures with that of the fish. In what respects are they similar? in what respects different? Why should they differ?

9. Where are the frogs' ears located? What do you think of the efficiency of an eardrum situated on the surface of the body? Why?

10. The frog has certain other responses. Try turning the jar or cage containing a frog around to face the frog in another direction. What happens? How do you explain this response on the basis of use? What other responses have you noticed?

Protection.

11. Has the frog an exoskeleton? Describe the color scheme of the frog and explain how it may be protective. Why are frogs brighter in spring than in fall?

12. Why do frogs usually live near water? Do they ever leave the vicinity of streams? If so, when?

Summary.

In what ways is the frog adapted to water life? In what ways to life on land? In what respects do toads differ from frogs?

The Frog's Mouth

Materials.

Preserved or freshly killed frogs in dishes or shallow pans of water; forceps and a bristle.

Observations.

Open the frog's mouth as widely as possible and, if necessary, insert a splinter to hold the jaws apart. Identify the following structures:--

Tongue. Draw it forward until the free end extends from the mouth and is outspread; observe its form, extent, and attachment.

Teeth. Find those on the jaws and on the roof of the mouth.

Nostrils. Push the bristle inward through a nostril to determine its direction and extent.

Vocal cords. These form a hard white mass in the floor of the mouth, well back behind the tongue.

Glottis, the slit inclosed between the vocal cords, opening into the trachea.

Esophagus, the passage to the stomach, at the posterior end of the mouth.

Eustachian tubes, small passages outward to the ears at the junction of the upper and lower jaws.

Questions.

1. Describe the probable action of the frog's tongue in catching a bug.

2. What advantage can you ascribe to the peculiar mode of attachment of the tongue?

3. Of what use is the notch in the inner end of the tongue? (Note its position when the tongue lies at rest in the mouth.)

4. If the frog chewed its food, how would the existing structure of the nostrils be very inconvenient?

5. Recall either the frog's habit of feeding or the structure of the nostrils. Do you think the nostrils are of much service in smelling? State the reason for your answer.

6. Of what use are the vocal cords and why are they so muscular? Consider their use in sound making and also their condition during swallowing.

7. Of what use are the teeth? Recall the form and use of the fish's teeth.

Suggested drawings.

a. The mouth, wide open and with tongue extended.

b. A diagram showing the path of air and of food through the frog's mouth.

The Organs of Digestion, Absorption, and Excretion

Materials.

Freshly killed or preserved frogs in dishes or shallow pans of water, forceps, and, if the pupil is to do any dissecting, scissors.

Directions and Observations.

The specimens may have been opened by the teacher, or may be dissected by the pupil as follows:--

Placing the frog on its back, with forceps firmly grasp the skin of the abdomen and the muscles beneath, just in front of the hind legs, and with the scissors cut straight forward in the middle line until the floor of the mouth is reached; this will separate the arms. Care must be taken not to cut too deeply, but this may be avoided by keeping the skin uplifted. Now cut sidewise in front of each hind leg in order that the body wall may be laid aside. Under the arms the heart will be seen; it will be studied as a part of the circulatory system.

* * * * *

Identify the following organs:--

Liver, the large red or brown mass, consisting of several divisions and lying close up under the arms.

Bile sac, small, green, and between the liver lobes.

Alimentary canal.

1. Mouth.

2. Esophagus.

3. Stomach, the elongated, light-colored, firm, and muscular portion.

4. Small intestine, a slender, more or less closely coiled, tubular portion.

5. Large intestine, a thin-walled enlargement at the posterior end of the canal.

6. Duodenum. This is a muscular portion of the small intestine immediately following the stomach, against which it is folded.

Pancreas, a yellowish, pulpy mass lying in the fold between the stomach and the duodenum.

Spleen, a dark red globule, usually smaller than a pea, lying nearly free among the folds of the small intestine.

Fat bodies, yellow fringe-like structures, sometimes found near the stomach.

Kidneys, a pair of elongated dark red organs, behind the spleen and against the back. Note their numerous blood vessels. Possibly the ureters, or urinal ducts, can be discovered and traced to their junction with the bladder, a clear membranous sac in the posterior extremity of the body cavity.

Peritoneum, a thin membrane lining the body cavity and attaching the vital organs to the backbone.

Note.--Specimens secured in late fall, winter, or early spring may contain, if female, a large number of dark-colored eggs; or if male, two white testes, located near the kidneys and similar to them in form, though smaller.

Questions.

1. Name the parts of the frog's alimentary canal.

2. Name the glands or organs which are accessory to the canal.

3. How long is the esophagus? How does the presence or absence of a neck affect the esophagus?

4. How does the thickness of the stomach wall compare with that of the intestine, and how do you account for the difference?

5. Measure the length of the trunk of the frog's body and that of the outstretched alimentary canal. How many times the length of the one is that of the other? How does this ratio compare with that of an herbivorous animal? (The sheep's food canal is about thirty-two times the length of its body.)

6. What is the color of the bile, as seen through the walls of the bile sac? This color is characteristic of carnivorous animals; in herbivorous forms it is yellow. Find its color in some omnivorous form, as man.

7. Name the organs concerned in excretion.

8. What holds the internal organs in place, and from what are they suspended?

9. The spleen is called a "ductless gland." Give its function, and explain why a duct is not necessary to it.

10. Since the frog swallows its food alive and entire, what work must the stomach do? What digestive organs would be absent from the mouth, or else poorly developed?

11. Since the frog is carnivorous, what digestive ferments are probably present, and what ones absent from the alimentary canal?

12. Fat bodies are largest in the fall, and are rarely found in the spring. How can you account for this?

13. When through with the general study of the alimentary canal, you may open the stomach by cutting it lengthwise. Describe the character of the stomach lining as to folds and villi, stating the advantage of each being present and the reason for the direction of the folds. Tell how the food is propelled onward through the alimentary canal. Give the scientific name for this action.

Suggested drawings.

a. The viscera (internal organs) undisturbed.

b. The alimentary canal extended.

c. The excretory system.

The Organs of Circulation and Respiration of the Frog

Materials.

The materials used in this exercise are the same as those used in the preceding exercise.

Observations.

The pupil should identify the following structures:--

Pericardium, a membrane that surrounds the heart and, in the case of the frog, separates the body cavity into two portions, the abdomen and the pericardial chamber.

Heart, lying between the shoulders and in front of the false diaphragm. It is conical in form and composed of three chambers.

Ventricle, the pink, conical, and muscular portion of the heart, pointing backward and outward.

Auricles, right and left. These are anterior and dorsal to the ventricle, thin, membranous, and dark-colored.

Arterial trunk, the single large blood vessel, usually empty of blood, and white. Note its origin and trace it as far as possible, at least until you see it divide to encircle the throat.

Sinus venosus, a large membranous sac dorsal to the heart and connected with the right auricle.

Lungs, two small oblong, pink, spongy sacs, lying between and behind the shoulders.

The pupil may also identify the following structures if a specimen is available which has the blood vessels injected.

Conus arteriosus, or "arterial trunk," a large artery passing obliquely forward from the ventricle, and dividing into three branches on each side.

Carotid arteries, the first branches of the conus, to the head and neck.

Aortas, the second branches of the conus, to the dorsal region.

Pulmocutaneous arteries, the third branches of the conus, to the lungs and skin.

Dorsal aorta, the large artery along the back, formed by the union of the two aortas.

Iliac, or femoral arteries, the two posterior divisions of the dorsal aorta, supplying the legs.

Posterior (ascending) vena cava, a large vein close to the dorsal aorta, passing forward from the kidneys.

Hepatic veins, large veins connecting the liver with the posterior vena cava.

Anterior (descending) venae cavae, large veins formed by the junction of the veins from the arm, neck, and head on the right and left sides.

Subclavian veins, from the arms.

Jugular veins, from the neck.

Questions.

1. Of how many chambers does the frog's heart consist? Name them, and describe them as to size, color, and structure.

2. Which chamber receives blood from the body, and which receives blood from the lungs?

3. Name the large arteries and give the regions which they supply. Name the large veins and give the regions from which they come.

4. Describe the lungs as to size, both when inflated and when uninflated. Describe their color; entirety or subdivision; texture.

5. In the inflated lung, notice the interior partitions or chambers, which are called vesicles. How do they affect the amount of surface exposed for gas exchange in breathing?

6. Measuring the lung collapsed and again when inflated, calculate its approximate volume in each case and state how much air it may take in during an inspiration.

7. Describe the diaphragm and state its probable use as a factor in respiration or as a partition.

Advanced questions.

8. Apparently the pure and the impure blood must commingle upon entering the single ventricle, but by a simple device this scarcely occurs. How would such commingling affect the purity of the blood as it reached the tissues, and hence affect waste removal, oxidation, body temperature, activity, and intelligence?

9. What are the chambers of the fish's heart? of the mammalian (human) heart? How may the heart of the frog be regarded as intermediate between these

others? How and where might the growth of a partition within it bring about the higher structure? (This actually occurs in certain reptiles.)

10. The lungs are said to be outgrowths of the alimentary canal. Explain how their connection would tend to show this.

Suggested drawings.

a. The heart and lungs in their normal position.

b. The circulatory system, as seen in the injected specimen.

c. A copy of the diagram or model of the circulatory system.

d. A diagram of the respiratory tract from the nostrils to the lungs, by arrows showing the course of the air.

The Nervous System of the Frog

Materials.

Specimens which have the brain exposed and other specimens whose viscera have been removed so that the spinal nerves can be seen; pans or shallow dishes of water and forceps.

Observations.

The nerve tissues are generally white in color unless they have been specially treated and stained. The pupil should identify the following structures:--

A. Brain, those enlargements of nerve tissue situated in the head and composed of four principal parts, as follows:--

1. Cerebral hemispheres, a pair of elongated lobes, the anterior enlargements.

2. Mid brain, or optic lobes, a pair of large ovoid structures, projecting diagonally forward and sidewise.

3. Cerebellum, a slender, transverse ridge, close behind the midbrain.

4. Medulla, the anterior end of the spinal cord, widest in front and containing a triangular depression.

(Frequently a pair of smaller enlargements is to be seen in front of the cerebral hemispheres; they are the olfactory lobes, and from them nerves pass forward to the nasal chamber.)

B. Spinal cord, extending along the spine, giving rise to nerves.

C. Spinal nerves, ten pairs of nerves which are connected with the cord through dorsal (sensory) and ventral (motor) roots, and which penetrate the body and its appendages. The first enters the neck; the second and third join and enter the arm; the fourth to sixth penetrate the skin and muscles of the trunk; the seventh to ninth join by a plexus to form the sciatic nerve which supplies the leg, and the tenth enters the posterior portion of the body.

Questions.

1. Which lobes of the brain are paired? Give at least two possible causes or reasons for their double structure.

2. The optic lobes are connected with the eyes. Compare their size with that of the other parts. Of how much use do you think they are to the frog?

3. If folds or convolutions in the surface of the cerebrum indicate intelligence, thoughtfulness, or mind, what do you infer as to the frog's mental condition and power to think?

4. Has the frog brain or "brains"? Explain.

5. Name several things done by an animal's brain.

6. Which of the spinal nerves are specially large? Why should they be large?

7. What advantages are there in the frog's having a dorsal nerve cord instead of a ventral one, as the earthworm has?

What would be the effect of cutting or breaking the dorsal root of a spinal nerve? the ventral root? the entire spinal cord? How do you explain the convulsions of the brainless (beheaded) chicken or frog?

8. What kinds of impulses originate outside of the nerve center, and what kinds in the center?

9. Sensations reach the cord and brain through the dorsal root of the spinal nerve. What kinds of messages travel through the ventral root, and in what direction do they go?

Suggested drawings.

a. The brain.

b. The spinal cord and its nerves.

c. The nervous system.

d. A diagram of the cord and its nerves, showing the kinds and the directions of the nerve impulses.

The Endoskeleton of the Frog

Materials.

Prepared frog skeletons mounted in glass-covered boxes or in other cases suitable for individual study; other vertebrate skeletons for reference.

Observations.

The pupil should examine his specimen and identify the following structures:--

A. Skull:--

1. Cranium, or brain case, the central and hinder portion.

2. Nasal bone, a triangular bone lying in front of each large opening, or eye orbit, and attached to the anterior end of the cranium.

3. Premaxillaries, a pair of small bones which form the tip of the nose.

4. Maxillary, a slender bone forming the side of the upper jaw.

5. Dentary, the bone of the lower jaw, corresponding to the maxillary.

6. Occipital foramen, the posterior opening or entrance into the cranium, normally covered dorsally by cartilage and most easily seen in a separate skull.

B. Vertebral Column:--

1. Cervical vertebra, the first vertebra, supporting the skull.

2. Dorso-lumbar vertebrae, vertebrae with small lateral processes.

3. Sacrum, the ninth vertebra, bearing extra long lateral processes or "arms."

4. Urostyle, the last vertebra, lying in the median line, a long, slender, blade-like bone, really formed by the fusion of several vertebrae.

C. Vertebra:--

1. Centrum, the solid, nearly circular portion.

2. Lateral process, one of the paired projections, extending outward on either side.

3. Neural spine, the single projection, extending toward the dorsal side.

4. Neural arch, formed largely by a connection between the lateral process and the neural spine. The arch above and the centrum below inclose the neural canal.

D. Pectoral, or Shoulder, Girdle:--

1. Sternum, the "breastbone," extending along the median ventral line.

2. Coracoid, a heavy bone extending from the sternum sidewise to support the arm.

3. Clavicle, the "collar bone," a light bone in front of the coracoid.

4. Scapula, the "shoulder blade," a broad, thin bone which arches around to the dorsal side.

E. Pelvic, or Hip, Girdle:--

1. Ilium, a long, slender, curved bone, with its mate uniting to form an inverted "wishbone"; it is joined to the sacrum.

2. Pelvis, the region of the socket at the junction of the ilia.

F. Fore Leg, or Arm:--

1. Humerus, the single bone of the upper arm.

2. Radius, the bone of the lower arm on the thumb side; in the frog united to the ulna.

3. Ulna, the bone of the forearm opposite the radius.

4. Carpals, small bones of the wrist.

5. Metacarpals, a single series of long bones forming the palm of the hand.

6. Phalanges, the bones of the fingers and thumb (singular phalanx).

G. Hind Leg:--

1. Femur, the thigh bone, next to the body.

2. Tibia, the larger bone of the lower leg on the inner side; in the frog united

to the fibula.

3. Fibula, the smaller bone of the lower leg.

4. Tarsals, the small bones of the instep.

5. Metatarsals, the long bones of the instep.

6. Phalanges, the bones of the toes.

Questions.

1. Whereabouts in the frog has nature made an attempt to inclose delicate or vital structures in bony cases?

2. Examining the inside of the mouth, find the teeth. Judging from their size and structure, of what use are these teeth?

3. What advantages can you see in having the arms and legs attached to girdles instead of having them fastened directly to the vertebral column? How has their development affected the shape of the trunk, as opposed to that of the fish?

4. Make a comparison of the two girdles as to their attachment to the spine and their consequent rigidity or freedom of movement.

5. How many vertebrae are there in the spinal column? What advantage can you see in having the column composed of many small vertebrae instead of a few large ones? Enumerate those having a special form or structure, and state the use of each.

6. On the vertebrae notice any irregularities, prominences, or roughenings. For what are such bones better adapted than smooth bones would be?

7. How is the frog's humping permitted? How are the urostyle and the pelvis connected, if at all? Has this any effect on motion? (See living frog.)

8. In parallel columns, keeping corresponding parts in a line, tabulate the

bones of the fore and hind limbs.

9. State how extra length has been attained in the hind leg, and give the purpose or the result of this lengthening.

10. In the forearm notice the fusing of the radius and ulna. How would these parts act in rotating the hand, as compared with your own, where they are free? Compare also the tibia and fibula. Where in the frog's leg is turning made possible by the use of parallel bones?

11. Compare the frog's hand and foot as to number of fingers and toes; as to length and spread. Have any fingers or toes only two phalanges?

12. How would an inner skeleton affect the growth and the size of an animal, as compared with an outer skeleton, like the clam's?

13. How would the lack of an outer skeleton influence sensitiveness, activity, and intelligence?

Suggested drawings.

a. The skull and trunk, with appendages of one side.

b. Each set of bones separately, as the skull, the column, etc.

Comparative Study of Amphibia

Materials.

Various amphibia, either dead or alive, such as newts, mud puppies (necturus), salamanders, and several species of frogs and toads, especially tree toads.

Observations.

Answer the following questions with respect to each animal:--

1. What regions of the body are present? What is the general shape and size

of the body?

2. For what kind of locomotion are the limbs fitted? How?

3. Have the hands and feet any special adaptations? If so, what are they and for what purpose?

4. With what organs does the animal breathe? If with gills, are they external or internal?

5. Judging from the specimen, what do you think is the habitat? Give reason for your answer.

Summary from the Comparative Study of Amphibia

1. Which of the amphibia in this study are fish-like in character? What are the fish-like characters? Do you think these fish-like amphibia are of a lower or higher type than the others? Give reasons for your answer.

2. Show how the variation in (a) the form of the body, (b) color patterns, and (c) the structure of the legs and arms of the amphibia are related to habitat and mode of life.

Amphibia: A General Review and Library Exercise

1. The usefulness of the American toad.

2. The breeding habits of the common frog; of the Surinam toad; of the obstetrical toad; of the "smith."

3. The development of the axolotl.

4. The habits of the tree frogs and their variable coloration.

5. The croaking of frogs and of toads.

6. The flying frog of Ceylon.

7. The distribution and habits of the hellbender; of the mud puppy; of the Congo "snake."

8. The general absence of the amphibia from arid regions.

9. Frog farming for city markets.

10. Protective devices of various amphibia.

11. Toads and warts.

12. The homing and water instincts of toads.

13. Hibernation, seasonal or prolonged, of toads and frogs.

14. The changes in the respiratory and circulatory system during metamorphosis.

15. Ontogeny and phylogeny, as illustrated by amphibia.

16. The structural defects which would prevent an ambitious frog from becoming a highly trained and skillful animal.

17. The classification of amphibia, with examples.

18. The kinds of frogs and toads found in the region where you live.

#C. STUDIES OF LIVING REPTILES#

#Showing Diversity of Adaptation among closely related Animals#

The Snake

1. What is the shape? What regions are present? How do you distinguish between the trunk and the tail? Has the tail any use? What are the advantages of a body without limbs?

2. Describe the snake's path in locomotion. How is locomotion accomplished?

Is the absence of limbs a hindrance to the animal in its locomotory activity?

3. Describe the appearance and movements of the tongue. Of what use are these movements of the tongue?

4. What evidence is there that the snake breathes with lungs?

5. What sense organs do you find? Compare the ears and eyes with those of the frog. What explains the peculiar staring appearance of the eyes?

6. What is the color scheme? Does it appear to be protective? If so, how?

7. Of what does the exoskeleton consist? How are the scales arranged? What variations in the size of the scales do you find? What special use have some of the scales? How fitted for this? How does a snake moult? Appearance before and after moulting.

The Florida Lizard

1. Describe the body as to regions and shape. Is the tail of use?

2. What are the lizard's locomotory abilities? How is it fitted for living in trees?

3. Feed the lizard flies or meal worms and describe its method of capturing them.

4. Compare the lizard's respiration and sense organs with those of the snake and frog.

5. What is the usual color of the lizard? What have you discovered about its power to change colors? How may this be of use to it?

6. Of what does the exoskeleton consist? Is the entire body covered? How are the scales arranged? How does the lizard moult?

The Turtle

1. In what important ways does the body of the turtle differ from those of the snake and lizard?

2. What methods of locomotion has the turtle? For which method is it best fitted? How? Why are its movements in water so much less clumsy than on land? (Compare weight on land with weight in water.) Compare the efficiency of the locomotion of a turtle with that of a lizard and suggest a reason for the difference.

3. What do turtles eat while in the laboratory? Since turtles have no teeth, how can they bite off their food?

4. What can you discover with respect to the respiration of the turtle? Does it breathe when under water?

5. What sense organs has a turtle? Which appears to be most highly developed? How do you know?

6. In what various ways is the turtle protected against enemies? (You should state several.) Are you afraid of a turtle? If so, why? Why does the turtle need more protection than the snake or lizard?

7. Describe the color pattern of the turtle you are studying. Is this arrangement of colors the same for others of the same species?

8. Describe the arrangement of the epidermal plates of the turtle's shell. Are the arrangement, number, and form of plates the same for all turtles of this species? Compare with the plates of other species. What variations do you find?

Reptiles: A General Review and Library Exercise

1. Characteristics of reptiles.

2. Orders of reptiles. Characteristics and examples of each order.

3. Poisonous snakes found in the United States. The poison fangs of a rattlesnake. Habits of the rattlesnake.

4. Cures for snake bites, fabled and real.

5. Snake charming.

6. Famous poisonous snakes and their habits.

7. The characteristics and habits of alligators.

8. The characteristics and habits of marine turtles. How they differ from pond turtles.

9. For what is each of the following noted? Where does it live? The terrapins? the horned toad? the Gila monster? the chameleon? the glass snake?

10. Reptiles of former (geological) times compared with those of the present.

11. Care of eggs and young. Swallowing of young by snakes.

12. Food of snakes. Defend the proposition that non-poisonous snakes are beneficial and should not be killed wantonly.

13. Investigate and write an account of the economic value of reptiles.

#D. STUDIES OF BIRDS#

#Vertebrates specialized for Flight#

The Living Pigeon

Materials.

Living pigeons in cages, and, if practicable, other birds such as chickens, canaries, and sparrows.

The Body.

1. What divisions of the body are present? Compare the relative size of these divisions with that of other vertebrates studied. Can you account for the great

increase in bulk of the trunk over the other divisions? How do you account for the longer neck?

Flight.

2. What is the shape of the body? How is the body made smooth? To what is the shape and smoothness adapted? (Recall the appearance of a plucked pigeon or chicken.)

3. Observe the size of the wings by measuring their width and spread (measured from tip to tip when outspread). Where on the body are they attached? How is this point of attachment advantageous? What is the shape of the upper surface of a wing when spread? of the lower surface? How is this shape advantageous in flight?

4. Where are the largest and strongest feathers? How are they arranged with respect to each other: (a) to prevent air passing through on the down stroke of the wing? (b) to permit folding?

5. What is the shape and width of the tail when outspread? when folded? In what way does the tail assist in flight?

Walking and Perching.

6. Identify the parts of the leg as in the frog. (See study of frog.) With what exoskeletal structures is each part covered? How many toes are there? Does the pigeon walk on its foot or on its toes? Where is the ankle joint?

7. The bird's legs are arranged to support the body. How is this brought about? Compare with the lizard and frog in this respect.

8. Explain how the legs are adapted (a) to preserve the balance of the body, and (b) for perching. Are they well or poorly adapted for locomotion? Explain. For what other purposes are the legs sometimes used?

Feeding and Drinking.

9. What is the form and character of the jaws of the pigeon (called the beak)?

Offer food, and watch the pigeons eat. How does a pigeon seize and swallow food? Does it chew the food? Locate the crop. If the pigeon chewed its food, would it have a crop? Why? How does a pigeon drink? How does the chicken drink?

Respiration.

10. What parts move in breathing?

11. Compare roughly the body temperature (by touch) of man, the pigeon, and the frog. Which has the highest temperature? which the lowest? How can you account for the higher temperature of the bird? (Which of the three must be able to endure long-sustained effort? What is the effect on your own body of long-continued effort?)

Sense Organs.

12. Identify the eyes, ears, and nostrils, and describe their location. What are the advantages in having the eardrum at the bottom of a canal? Is there any disadvantage? What is the probable function of the inner thin eyelid? How does it work?

13. Investigate the power of response of the special senses. Which seems to be the most alert? State the experiments which you used and your reasons for your answers to this question. Which senses are most serviceable in finding food? in protecting against enemies?

Protection.

14. In what different ways are feathers protective to the bird? Study the arrangement of the feathers to find how protection is increased in this way. How do the feathers and parts of feathers which lie next the body differ from those which are on the outer surface? What parts of the body are unprotected by feathers? How are these protected, if at all?

15. Note the flexibility of the neck. Through what part of the arc of a circle can the pigeon turn its head? How is this especially important to birds?

16. What means of defense has the pigeon when attacked?

Summary.

1. Make a list of the important adaptations of the pigeon, (a) to flight, (b) to feeding, (c) to perching, and (d) to protection.

2. Which is the most highly specialized, the fish, frog, snake, or bird? Give reasons for your answers.

3. In what various ways are feathers of use to the bird?

Suggested drawings.

a. Side view of the head.

b. A foot.

c. A wing.

d. Side view of the pigeon.

Supplementary Studies of the Pigeon

These topics and questions should be answered from experience with pigeons and observations of their habits and behavior, and from books to be obtained from the school and public libraries.

1. The homing instinct of the common blue pigeon and of carrier pigeons.

2. Nesting habits, number of broods and number of young in each brood, feeding the young. Why is this method of feeding the young necessary?

3. Varieties or breeds of pigeons. How the various varieties were produced; how they are kept true; reversions of type.

4. Darwin's experiments with the pigeons; object of the experiment. Whitman's experiments.

The Exoskeleton of the Bird: Plumage

Materials.

Living birds, as perhaps pigeons, or mounted or stuffed skins, separate feathers, portions of feathers, microscopes.

Observations.

The pupils should study the arrangement of the feathers and their variations in form and size, and should identify the following principal kinds:--

Contour feathers, those feathers, generally broad, which cover the body, giving to it its outline and color.

Coverts, those feathers which cover joints, such as the joints of the wing and tail.

Primaries or pinions, the long stiff feathers of the outside of the wing, used in sustaining the bird in flight.

Secondaries, the shorter, more symmetrical feathers lying next and over the primaries.

Down, soft feathers found on young birds and next to the skin on some adults.

Thread feathers, best seen about the eyes, ears, and beak.

Quill, the bare stiff portion, one end of which is inserted in the skin. Examine its internal structure.

Vane, the broad expanded portion, the part ordinarily seen on the bird.

Shaft, the mid-rib of the vane.

Barbs, the delicate outgrowths of the shaft making up the vane.

Barbules, the subdivisions of the barbs, some of which are provided with hooklets. These may well be studied microscopically.

Questions.

1. How is the plumage arranged to offer least resistance to the air in flight? How does a bird sit when exposed to the wind?

2. In a column write the names of the parts of a feather, and opposite each part state its particular use.

3. Of what use are the hooklets on the barbules?

4. State and explain the peculiar position of the shaft of the primary feather; of that of a contour feather.

The plumage of many birds contains all stages of feather specialization, from down to pinion. Find as many of them as you can on your specimen.

5. What reason can you assign for the fluffy base and the compact end of the contour feather?

6. How much of the feather of a full-grown bird seems to be supplied with blood vessels? How does this bloodless or full-blooded condition affect the weight of the feather; of the plumage; of the bird?

7. How would the above condition affect the temperature of the blood and of the body? Would it tend to cool the animal or to conserve its heat?

8. Show how the air lying motionless amongst the plumage may serve the same purpose as that in the packing of a fireless cooker or a "thermos bottle."

9. How would the above condition affect the growth and repair of the feather? What connection has it with moulting?

10. What kind of feathers forms most birds' "baby clothes"? What kind forms the adults' "underclothes"?

Suggested drawings.

a. A typical feather.

b. A feather of each kind.

c. A bird with wings outspread, showing positions of feathers.

d. The minute structure of a feather.

Birds and Migration

To illustrate Distribution

Work in the Laboratory

Materials.

Bird skins or mounted birds, at least one representative of each order and, better still, of each family of the birds which pass through or remain in your neighborhood; specimens in a museum may also be used. Some guide to the identification of birds, as Walter's "Wild Birds in City Parks" or Reed's "Bird Guide." A good diagram of a bird.

Directions.

Your object here is to familiarize yourself with the appearance of birds of different types, and with the terms used in describing them. Study first a diagram of a bird and learn the terms and their applications.

An important item in the description of a bird is its length. This is obtained by measuring from the tip of the beak, over the curve of the head, to the end of the tail. This measures a curved line and gives a greater length for a bird than the straight line you would naturally estimate. Train your eye by watching house sparrows (so-called English sparrows) and fixing their length as a unit. They are called six inches long, and in the field other birds may be measured by them. The robin is ten inches long, and may be used to measure the larger birds.

In describing the colors of birds, great discussions often arise because pupils do not use terms correctly. Consult Chapman's "Birds of Eastern North America" for a color key or chart, and train yourself to observe colors carefully and name them correctly. You will find more colors among birds than are given there, but the chart will help you a great deal.

Study in the laboratory as many birds as you can. Try to get one of each order at least and, among the perching birds, one of each family. The answers to the following questions should be recorded upon the blank outlines of birds, or else in the form of a table.

Questions.

1. How long is the bird?

2. What is the general color of the upper surface; of the lower?

3. What are the markings or peculiarities of the head, if there are any?

4. Note any peculiarities of the tail, as to shape, length, or color, if there are any.

5. If the wings are not like the back, note their color, and, if there are wing bars, note their number and color.

6. What are the markings on the breast, if there are any?

7. Note any other markings, as rump spot, etc.

8. What sort of beak has the bird?

9. What sort of feet has it?

10. Identify the bird, using any key or guide you have. Do not ask any one to help you until you have exhausted your own resources.

Comparative Study of Birds

1. In what ways do the feet of birds vary? Give examples to illustrate your answer. What are the principal results of these variations?

2. In what ways do the beaks of birds vary? Give examples of the variations and of the special uses of the beaks.

3. How do water birds differ from land birds; flesh-eating birds from seed eaters; insect-eating birds from seed eaters; shore and swamp birds from land birds?

Work in the Field

Materials.

(1) Birds in the field, field or opera glasses, and bird guides. (2) Some extra time, as field work is rather slow. (3) Considerable energy, as birds rise early and may be up and away before the usual hour for your appearance.

Directions.

The object of this work is to become acquainted with the living bird, to learn not only its name, but also some of its ways. You will need to spend time to do this, and as a rule the more time you spend the more you will see. Every time you go out after birds, record at the time every kind of bird you see, so that at the end of the season you will know not only when each kind of bird came, but also how long it staid. When you see a new bird, record immediately its colors, markings, actions, notes, and anything else which may help you later to identify it. Do not trust to memory nor to the inspiration of the classroom. After weeks of observation, write the following summary.

Summary of the Results of Field Study of Birds

1. Over what length of time have your observations extended? Where have you studied? What have you found to be the best conditions for studying birds? How many birds have you identified?

2. When in the year do birds migrate; when in the twenty-four hours?

3. In spring migration which birds come first; which come last? What reason is there for this order?

4. What may retard migration? What may hasten it?

5. What could prevent certain birds from ever coming here, or, if they did come, from staying?

6. Name some birds which stay here permanently; some which come only for the winter; some which come for the summers; some which merely pass through, going and coming.

7. Can you see anything which may determine whether a bird will nest here or farther north? If so, what is it?

8. Why is the house sparrow so successful?

9. Why are blue jays so nomadic in winter?

10. What months do the herring gulls stay here? When do they leave? Where do they go when they leave? What do they do while they are gone? When do they return? What is their economic value?

11. How many birds' nests have you seen this spring? To what kinds of birds did they belong? If you have been able to study one in particular, give its history as far as you know it.

12. Tell what you have learned by your own observation this spring concerning the kinds of food birds eat, and their methods of obtaining food.

13. What bird songs have you learned to know? When do these birds sing most? Does a bird have more than one song?

14. What birds have you seen near your home? What attached them to the vicinity? How might you attract more birds?

Birds; Review and Library Exercise

1. What are the distinguishing characteristics of birds?

2. Give the orders of birds, with the characteristics of each order and an example of each.

3. Define and give illustrations of the meaning of the expressions: "land birds;" "water birds;" "shore birds;" "swamp birds;" "scavengers;" "policemen of the air."

4. Discuss the temperature of birds, the amount of energy they show, and the oxygen they use.

5. Beaks of birds.

6. Feet of birds.

7. The crop and gizzard. Why absent in many birds?

8. What is there in a bird's construction which enables it to twist its head so far around? What are the advantages in being able to do so?

9. Account for the oiliness of plumage of water birds; for the complete feathering of legs and feet in some forms.

10. Define moulting and discuss its advantages and disadvantages.

11. Give reasons why you would not expect a bird to hibernate.

12. Name two or more kinds of birds which show "recognition marks." What sort of birds would you expect to find with such marks? Why?

13. Is a bird's egg alive when it is laid? Why does it have so much food stored in it? What direct interest have we in this fact?

14. How are the eggs of various birds protected while they are developing?

Note.--Do not be satisfied with only one way. A bird seldom is.

15. Compare praecocial and altricial birds as to their stage of development when hatched; the location and character of the nest; the care given by the parents; the singing habit of the parents; the success of the type.

16. Name at least three insect-eating birds and tell how many insects it is estimated each will destroy in a day. Effect in a garden?

17. If you had an orchard to protect from insects would you spray it with poison, or would you police it with birds?

Note.--Consider both sides. There is much to be said on each.

If you decided that you needed more birds, how would you get them?

18. Suppose you had a city lot in the suburbs, 50 feet wide by 200 feet deep, with a house covering the first 50 feet. Make a plan of the back yard to show what you could do to attract the birds to it in the summer; in the winter. Remember that birds must have protection against enemies as well as against changes in weather, etc.

19. For what purposes are birds killed by man? Which of these do you consider legitimate? Which birds may not legally be killed at any time of the year in this state? Which may at certain seasons? Which may at any time? When should an "open season" be permitted? How long should it last?

20. Give a short biography of Audubon. Describe the purpose of the Audubon Society and some of the work accomplished by it.

21. When were the house sparrows, or, as we call them, the English sparrows, introduced into this country? Where? Why? Have they proved a success from our standpoint? From their own? Why have they increased so enormously? What part of the country is still free from them? Why? Give the reasons for and the methods of fighting English sparrows.

22. The history of the passenger pigeon.

23. What has the quail or bobwhite to do with our food supply?

24. Poultry keeping.

25. Name three bird magazines and give a characteristic of each one.

26. Fossil birds and the light they shed on the probable ancestry of birds.

Study of the Migration of Animals in General

The migration of birds is only one case of a phenomenon which is comparatively common. From your text and reference books find other examples of migration. What are the causes which make animals migrate? What methods do various animals use? What are some results of these migrations? Summarize your study in the following thesis:--

The distribution of animals.

1. The necessity for this distribution.

2. Methods of distribution--voluntary--involuntary.

Note.--See in review the methods used to spread corals, hydroids, and other sedentary forms, starfish, clams, etc., as well as those used by the various vertebrates.

3. Time when migration occurs. Consider here the young of most animals, and the movements of many birds, as well as movements caused by some accidental occurrence.

4. Distance that animals move from the place of their birth.

5. Factors which determine the routes of distribution.

6. Factors which limit distribution.

Migration maps.

1. A map to show the migration route of the birds of your region.

2. A map to show the migration of the potato beetle (or English sparrow or any other animal the extension of whose range has been studied).

3. A map of the world, showing the zonal areas.

4. A map of North America, showing the distribution of the ungulates, with the boundaries and barriers marked.

5. A map of North America, showing the distribution of the fur-bearing animals.

6. A map of the world, showing the distribution of the human races.

#E. STUDIES OF MAMMALS#

#To illustrate Man's Relation to Other Animals; the Connection between Mode of Life and Structure#

The Rabbit

The effects of domestication upon an animal. A burrowing type of rodent.

Materials.

Living rabbits--young rabbits are more desirable for laboratory study.

Observations.

The Body.

1. What divisions of the body are there? Compare the length and use of the neck with that of the pigeon. Describe the character and length of the tail. What use has it, if any?

Locomotion.

2. What methods of locomotion has the rabbit? Which is most commonly

used?

3. Study the limbs, and find the ankle and wrist. Does the rabbit walk on its foot or its toes? Note the number and character of the toes and fingers and their claws. What is the effect of the nonretractile power of the claws upon the uses to which claws can be put?

4. What uses have the fore or the hind limbs other than locomotion? Explain how the usual sitting or resting posture is advantageous for quick locomotion.

5. By means of printer's or writing ink smeared over the soles of the feet, and a long piece of clean white paper get prints of a rabbit's tracks as it hops over the paper. Explain how the peculiar formation of the tracks occurs.

Feeding.

6. Offer a rabbit various kinds of food. How does it test the food before eating? Does the rabbit ever use its forelimbs to assist in feeding? If so, when and how? How is the food eaten? How are the teeth fitted for the rabbit's method of feeding? Does your rabbit drink? If so, how?

Respiration.

7. What movements of the body are concerned in breathing?

8. Compare the frequency of the breathing movements with your own. Can you distinguish the heart beats? If so, how and where? Count them.

Sense Organs.

9. Note the position of the eyes, shape of their surface, shape and size of the pupil. What external protecting structures are present,--such as eyelids, eyelashes, eyebrows? Study the movements of the eyelids. Is there a third eyelid as in the bird? Compare the use of the neck in enlarging the range of vision with that of the bird. Is the rabbit's range of vision greater or less than your own?

10. What is the location of the nostrils? What advantages are gained by the

prolongation of the face forward? Note any peculiarity in the form or movement of the nostrils. How do you explain these movements? Are the nostrils more or less useful than those of other vertebrates you have studied? How?

11. Note the shape, size, and position of the external ears and explain what relation, if any, these characters of the ears have to hearing. Note the various movements of the ears and the reasons for these movements.

12. What special organs for touch has the rabbit? Under what circumstances are these of use? What parts of the body are most sensitive to touch?

Protection.

13. What home-making habits do you observe in the rabbit? What habits relating to secrecy, comfort, and safety, do you observe?

14. Study the fur and hair of the rabbit. How do you distinguish between fur and hair? What variations do you find in the fur and hair? What parts of the body are uncovered? Why? What are the various functions of the fur and hair covering of rabbit?

15. Have rabbits any means of defense or offense? Explain.

Social Habits.

16. Notice and describe anything in the behavior of the rabbits which may be classed as social,--such as play, fondness for company, display of affection, homing instincts, care of young, etc.

Supplementary Study of Wild Rabbits

If you cannot answer these questions from observations of wild rabbits, the answers may be obtained by reading some good natural history. Ernest Thompson Seton's story of a rabbit's life is good for the purpose.

1. What method of locomotion is more highly developed in wild rabbits than in domestic rabbits? Why?

2. When do wild rabbits do their feeding? Why? In what ways do they sometimes do damage in feeding?

3. What senses will probably be more alert than those of the domestic rabbit? Why?

4. Where do wild rabbits usually make their homes? Why? How do they guard against being cornered in their homes?

5. In what ways do they guard against surprise when feeding? What are the principal enemies of rabbits? What devices do they employ to escape enemies when pursued by them? What is thumping? When used?

6. How many young rabbits are usually produced at one time? How many litters in a season? How long does it take a young rabbit to mature?

7. How are the young of rabbits guarded against danger from enemies and weather? What are the various causes that tend to keep down the numbers of rabbits?

8. Give an account of the plagues of rabbits in Colorado and Australia, including the reasons for the great increase in numbers and the methods used to destroy the rabbits.

Summary of the Study of Rabbits

1. What has been the general effect of domestication upon rabbits?

2. What are the most important characters and habits that fit the wild rabbit for its life?

The Guinea Pig or White Rat

Materials.

Living animals.

Observations.

1. What regions of the body can you identify? What is the relative length of the neck, ears, legs? What about the tail?

2. Describe the color scheme of the animal. Is it protective or the result of breeding? What is the character of the covering?

3. Describe the method and rate of locomotion. Would this method of locomotion enable the animal to escape from enemies (e.g. dogs)?

4. What is the shape of the foot? What is the shape and length of the claws? For what are they adapted?

5. What sounds do the animals make?

6. What is the appearance and shape of the eye? What is the color of the eye?

7. State the size and shape of the external ear. What movements are characteristic?

8. What motions of the nostrils do you see?

9. Feed the animal various kinds of food. How does it eat? State any facts you observed, to show that it has or has not a choice as to food.

10. Watch the animal for some time to determine its mental characteristics. Is it alert? curious? timid? Does it show much intelligence? affection?

What is the relation between mental development and success in the struggle for existence?

Summary.

1. What are the general characteristics of the animal?

2. To what kind of life is it adapted?

3. What are some of the characteristics that make the animal a good pet?

The Squirrel

Materials.

Living specimens in cages, mounted specimens, pictures, charts, lantern slides, etc.

Directions.

Before taking up the study of the squirrel in the laboratory a trip should be made to some park or wooded region and the habits of squirrels noted. Take your camera and try to take some snapshots. After the laboratory exercise visit some museum or zoological garden and study the relatives of the squirrel.

Observations based upon field work.

1. What different postures does the squirrel assume?

2. What does it do when frightened?

3. What use have the forelegs other than locomotion?

4. How does a squirrel go up a tree? down? from branch to branch? State all the forms of locomotion you have noticed.

5. What is the appearance of the tail? What is the position of the tail when the squirrel is sitting? running? on a branch? Describe any motions of the tail you noticed. Is there anything expressed by these motions or are they without meaning?

6. Tempt the squirrel with some nuts. State the evidence that leads you to think that the squirrel is alert, timid, curious. Do you think the squirrel acts most from instinct or as the result of intelligence?

7. In what various ways does a squirrel attempt to escape notice? What does it do when you chase it?

Observations in the laboratory.

1. What divisions are there to the body? What is the length of the neck? the length and appearance of the tail?

2. What is the relative length of the legs as compared with the body? How does the length of the front and hind legs compare?

3. Does the animal walk on its toes or on the sole of its foot? How many toes on each foot? What is the length of the claws? For what could they be used?

4. Offer the squirrel various kinds of food and see if it has a choice. Describe its methods of eating.

5. Note the position of the eyes, the shape of their surface, and the shape and size of the pupil. How many eyelids do you notice? Why do the squirrel's eyes appear so "bright"? Are eyebrows, eyelashes, or tear glands present?

6. Note the size, shape, and appearance of the squirrel's external ears.

7. What movements of the nostrils do you notice? For what does a squirrel chiefly use his nostrils? What explanation can you suggest for the nostrils, eyes, and ears having the same relative position in all vertebrates?

8. How does a squirrel protect itself?

9. Smear the feet of a squirrel with ink and allow it to run over a roll of clean paper as in the case of the rabbit. How do its tracks differ from those of the rabbit?

Summary.

1. What are the general characteristics of the squirrel?

2. To what kind of life is it adapted?

3. What adaptations has the squirrel to protect it from its enemies?

4. What characteristics make the squirrel a good pet? What objections to it?

Library Exercise on Rodents

1. General characteristics and examples of rodents. The teeth of rodents.

2. Show how variation in habitat depends upon structure among rodents by comparing, for example, squirrels, beavers, and woodchucks.

3. Variations in the tails of rodents. What are the causes of this variation?

4. Pocket gophers and their economic relations.

5. Species of mice. Their habits.

6. The dancing mouse.

7. Damage by mice. Plagues of field mice in Nevada. Method of extermination.

8. Habits and kinds of rats.

9. Economic importance of rats. Methods of extermination.

10. Rats and the bubonic plague.

11. Squirrels, kinds and habits.

12. The economic value of rabbits.

13. The groundhog myth. Habits of woodchucks.

14. The beaver--their habits and sagacity. Methods of trapping them.

15. Prairie dogs--their habits and economic importance. How exterminated?

16. What are porcupines?

17. Variation in the homes among rodents. Usual means of defense.

18. Make a list of rodents in a column, and in another column opposite each name write the various ways the animal is of economic importance. Sum up with a statement showing the most important ways rodents are of value to man and harmful to man.

19. Defend the proposition that rodents are on the whole harmful animals and should be exterminated.

20. How some rodents contribute to the science of medicine, more especially to bacteriology.

The Cat or Dog--Carnivora

Materials.

Living specimens of cats or dogs. Pictures, books, lantern slides, etc. Supplement the laboratory study with trips to museums and zoological gardens to observe the relatives of the cat.

Definitions.

Carnivora. An order of mammals, chiefly flesh-eating, with claws and well-developed canine teeth.

Carnivorous, flesh-eating.

Herbivorous, plant-eating.

Omnivorous, eating both plants and animal food.

Digitigrade, walking on the toes.

Plantigrade, walking on the soles of the feet.

Vibrissae, long hairs on the face--"whiskers."

Observations.

1. Into what regions is the body divided?

2. What is the shape of the head and the length of the neck?

3. Are the legs relatively long or short? How do the front and hind legs compare in length? How many toes on each foot? Is the cat digitigrade or plantigrade?

4. How many pads on the sole of the foot? What use can you suggest for these structures? What is the size and shape of the claws? Are they retractile or nonretractile? For what purposes may the claws be used?

5. Describe the tail as to length and appearance. Movements.

6. What is the size and appearance of the external ears? What movement do you notice?

7. Are the eyes large or small? What eyelids can you find? What other accessory structures? What is the shape and direction of the pupil?

8. What other sensory structures do you find? What is their function?

9. Watch the animal eat. Does it chew or "fletcherize" its food? What teeth seem well developed? Is the movement of the jaws simply up and down, or is there a lateral movement as well?

10. Try to find out some of the mental characteristics of the animal, i.e. is it sluggish or active? Is it alert? Does it show curiosity? fear? What evidence of intelligence?

Supplementary studies.

a. Smear the feet of a cat with ink and allow it to run on a sheet of clean paper. Make a diagram to show tracks. Do the same in case of a dog. How do these tracks differ? Why?

b. What is the difference between a cat and a dog as to the manner of eating a bone?

c. As you see dogs and cats outside do you see any evidence in either case of a tendency to gather in packs (gregariousness)?

d. What different emotions are expressed by a dog's tail? a cat's tail?

e. What sounds do cats and dogs make? Significance?

f. Contrast the sleeping habits of cats and dogs.

g. How large is the litter in case of dogs and cats? Condition of young at birth? How long before the eyes of the young are open? Care of young.

Summary.

To what kind of life does a cat or dog seem best adapted: (a) as to food? (b) protection from enemies?

Carnivora; Review and Library Exercise

Characteristics.

1. The general characters of carnivora.

2. Five important families. The characteristics and examples of each family.

Morphology and physiology.

3. The dentition of the cat, the dog, and the bear. Variation in the "chewing teeth."

4. Three types of paired appendages among carnivora. Relation to habitat.

5. The difference in structure and use of the posterior legs of the seal and walrus.

6. The alimentary canal of a cat and rabbit compared.

7. The tongue of cats and dogs contrasted as to structure and use.

Economics.

8. The difference between hair and fur.

9. The fur-bearing carnivora. Families, and habitat.

10. Trapping.

11. Game laws and game wardens. Hunters' licenses.

12. Hunting big game.

13. Carnivora harmful to man.

14. Carnivora useful to man.

15. Carnivora as pets.

16. Chief types or breeds of domesticated dogs. Characteristics. Special value of each.

17. Chief types or breeds of domesticated cats.

18. Dogs as burden bearers.

Natural History.

19. Distribution and range of carnivora. Carnivora of the United States.

20. Winter habits among carnivora.

21. Food of carnivora. Various methods of obtaining it.

22. The hunting habits of the dog and cat family.

23. The habits and distribution of the raccoons.

24. The color schemes of the more important families of the carnivora.

25. Seasonal variation in color.

26. Distribution and habits of the ferrets and weasels.

27. How carnivora protect themselves from enemies. Which carnivora have been most successful in resisting man's advance?

28. Peculiar and interesting carnivora to be seen in museums and zoological gardens.

29. Origin of the domestic dog.

30. Intelligence of dogs.

31. Fox-hunting.

32. Coyotes and their relation to stock raising, etc.

The Ungulates

Materials.

Pictures, charts, lantern slides, and books showing cow, sheep, hog, goat, horse, etc. Diagram of skeleton.

Directions.

Since it is impossible to have living ungulates in the laboratory, this study should be supplemented by trips to a museum and to a zoological garden. Observe also such hoofed animals as may be common in your neighborhood. Use your camera and make "snapshots," showing characteristic attitudes of these animals.

Definitions.

Ungulates, an order of mammals characterized by the possession of hoofs.

Ruminant, a division of ungulates, which "chew the cud."

Perissodactyl, a division of ungulates with an odd number of toes.

Artiodactyl, a division of ungulates with an even number of toes.

Carnivorous, flesh-eating.

Herbivorous, plant-eating.

Omnivorous, eating both plant and animal food.

Observations in the laboratory.

Note.--Answer the following questions for one or more of the following: The cow, sheep, goat, hog, and horse. If desired, the questions may be answered in the form of a table.

1. What is the relative length of the neck? What is its direction with reference to the body? Of what importance is this length and direction?

2. What is the length and appearance of the tail? What is its use?

3. What is the relative length of the legs? Locate the heel, knee, and elbow. (Reference should be made to a diagram of a skeleton.) When the leg is long, in which bone is this lengthening accomplished (compared with human skeleton)?

4. How many toes on each foot? Is the animal an artiodactyl or a perissodactyl? Is it plantigrade or digitigrade?

5. What is the relative size and position of the ears (external ear)?

6. What is the relative size and position of the eyes? nostrils?

7. Are horns of any kind present? If so, note the size, shape, and direction. Are they present in both sexes? If not, in which one? If in both, note any differences distinguishing the sexes.

Suggested drawings.

a. Head, side view.

b. Entire animal, side view.

Observations in the field or at home.

1. Note how the animal uses its lips, tongue, and teeth in feeding. Is it a ruminant?

2. In what order does the animal use its feet? Look up the definition of walk, run, gallop, canter, trot, lope, single foot, pace. Which of these forms of locomotion are optional with the animal?

3. Describe the process when the animal lies down and gets up.

4. Describe the covering of the animal, noting its length, fineness, etc. What variations in different regions of the body?

5. Is the animal alert or sluggish? Upon what senses does it most depend? What mental characteristics are most marked, e.g.curiosity, fear, suspicion.

6. Note any movements of the ear. What is gained by these movements?

7. What is the position of the eye? What is the shape and direction of the pupil? Reason?

8. What means has the animal for getting away from its enemies.

Observations based upon museum trip or natural history.

1. Identify as many ungulates as you can; for example, buffalo, musk ox, big-horn sheep, Rocky Mountain goat, chamois, antelope, giraffe, red deer, elk, moose, reindeer, wild boar, peccary, rhinoceros, zebra, hippopotamus.

2. Answer the following questions about each:--

a. What is the family, scientific name?

b. What is the size of the animal? the relative length of the hind and fore legs? the relative length of the neck?

c. What is the nature of the covering of the animal?

d. Are any horns developed? If present, what is their size, shape, direction, and appearance?

e. What is the habitat of the animal? its distribution and social life?

Summary.

In a short thesis summarize the facts you have found out about ungulates, using the following outline:--

1. Why called ungulates? Variation in number of toes.

2. General fact about the food of ungulates. The two divisions.

3. The general adaptations for protection.

4. The social life of the ungulates.

5. The native ungulates of the United States.

6. Commercial uses and value.

Ungulates: Review and Library Exercise

Characteristics.

1. Classification of ungulates based upon number of toes, kind of horns, "chewing the cud," etc. Some of the more important families with examples.

Morphology and physiology.

2. The variation in the number and kinds of teeth. The dentition (or dental formula) of horse and cow.

3. The various types of horns. Shedding of horns and sexual variation.

4. The structure and function of the stomach of a ruminant. Meaning of the cud-chewing habit.

5. The structure of the stomach of a camel.

Economics.

6. Ungulates which have been domesticated.

7. Breeds of cattle--their distinguishing marks and valuable points.

8. Breeds of horses--their distinguishing marks and valuable points.

9. Breeds of sheep--their distinguishing marks and valuable points.

10. Breeds of hogs--their distinguishing marks and valuable points.

11. Angora goats.

12. The making of butter and cheese. Kinds of cheese. Substitutes for butter.

13. The packing industry. Ungulates useful as food.

14. The various methods of preserving meat.

15. Cattle ranches and "round-ups." Free cattle in winter.

16. Cattle raising in your state; in other countries.

17. Transportation of cattle. Stock cars, care and feeding.

18. Useful products derived from ungulates.

19. Tanning. Varieties and use of leather.

20. Diagrams showing chief cuts of meats.

21. Sheep husbandry. Shearing.

22. Ungulates as beasts of burden. Advantages and disadvantages.

23. Government inspection--quarantine.

Natural History.

24. Geographical distribution of ungulates. Habitat and range.

25. Native ungulates of North America.

26. How, when, and by whom cattle and horses were introduced into America.

27. The geological history of the horse.

28. The story of the buffalo.

29. Deer hunting.

30. Methods of protection from enemies among ungulates.

31. Breeding habits and care of young in case of ungulates.

32. Intelligence in the case of horses.

33. Strange and peculiar ungulates to be seen in museums and zoological

gardens.

The Horse

The pupil is expected to study carefully the account of Eohippus or Hyracotherium in his text or any other available reference book, and to supplement that work and this brief sketch with original observations upon horses on the street, at a local store, or wherever possible or convenient.

From the early horses which migrated from North America there arose in Asia and Africa the ass, famous in the history of early civilization and still used in some localities as beasts of burden or for the breeding of mules, which are the crosses between ass and horse. There also arose the zebra and the most primitive of modern horses, Przewalskii's horse, a wild pony of western China, about forty inches high and almost identical with the drawings of the horse made by early man, 30,000 years ago. Doubtless the modern ponies of Ireland, Iceland, and Shetland are descendants of the original Przewalskii type and not, as is often claimed, true horses stunted by rigors of climate and scant fare.

The horse is characterized largely by the presence of a lock of hair between the ears, a full mane and tail, small ears, large hoofs, and peculiar neigh. The ass has no forelock, a scanty mane and tail, long ears, small hoofs, and a distinct bray.

By means of various crusades and raids, the modern horse was introduced into Europe from Asia, where it is clearly traced in history to the reign of King Solomon. Here, in Europe, because of local conditions and demands, it assumed differing type forms. The roadster type is closest to the Arabian in character. The draft or heavy type was bred in western Europe when heavy armor came into use for rider and horse, and the coach or carriage type was developed when armor was abandoned for gunpowder. Finally explorers and colonists brought the horse back to America, its original home.

The various types and varieties may be briefly described.

A. The draft type has short legs, short neck, large round body, and ranges in weight from 1400 pounds to 2000 pounds.

Varieties:--

1. Percheron: generally about 1700 pounds in weight, 16 hands (64 inches) high, gray or black, blocky body, steep rump, clean legs, and quick action.

2. Shire: generally about 1800 pounds in weight, 17 hands high, bay or brown, white marked feet and face, hairy legs and feet, and slow action.

3. Belgian: generally about 1800 pounds in weight, 16 hands high, chestnut or roan in color, compact body, short, steep rump, and small feet.

B. The coach or carriage type has legs and neck of medium length, a body full-chested but not blocky, and a weight varying from 1150 pounds to 1400 pounds.

Varieties:--

1. Hackney: generally of full, broad, powerful body, short legs and back, high action and high carriage of neck and head, bay or chestnut in color, 15 hands high, and 1400 pounds in weight.

2. Coach: generally lighter than the Hackney, with longer legs and long stride; height, 16 hands; weight, 1300 pounds.

3. Cleveland bay: averaging 16-1/2 hands in height, 1350 pounds in weight, high, broad hips, strong action, and bay color.

C. The roadster type is long and lean of limb and body, and averages about 1100 pounds in weight.

Varieties:--

1. Thoroughbred: of small head, long neck, level back, of variable color, 14-1/2--16-1/2 hands high, about 1000 pounds in weight.

2. American saddle: an American production; not a distinct breed, but a roadster of high quality.

3. American trotter: a superior type of good speed. The off forefoot and the nigh hind foot act together, the nigh fore and the off hind feet together, giving a two-beat gait.

4. Pacer: similar to the trotter, but using both off feet and both nigh feet together, giving a swinging gait.

The horse is very similar to man in its physical and mental character, being subject to the same ailments and treatment and having the same impulses of affection, hatred, fear, jealousy, obedience, willfulness, memory, and perhaps reason. It is of all animals most careful in its eating and drinking; because its stomach is small, the food should not be bulky but concentrated, grain forming a goodly portion of the ration.

Perhaps the most important point in the structure of the horse is the form of the leg and foot. The shoulder should slope slightly backward and the pastern joint, immediately above the hoof, slightly backward. The hips, or "quarters," should slope downward somewhat, and the hock should be comparatively wide to afford ample leverage for the pulling muscles. The legs should be straight as pillars when seen from front or rear. The outer walls of the hoof support most of the weight though the frog should normally touch the ground. In nature the hoof wears away properly of itself, but the shod hoof needs regular trimming attention, while the frog must not be trimmed, for it is the soft growing part that nourishes the hoof. In this treatment the foot is comparable with the human finger and finger nail.

Observations.

If access to a living animal is impossible or inconvenient, the pupil may use reference book or pictures for most of these points. A measuring tape or ruler should be at hand, and the assistance of an experienced person is a valuable aid. If several horses are studied, they should be distinguished by name or number.

Record the color, condition, weight, and height of the horse at the shoulder. (Height is given in "hands," a hand being the breadth of the palm, or 4 inches.) Note the slope of the shoulder, of the back and the hips, the general form of the head and neck, and the facial expression. Find the chestnuts, warty growths on the inside of each leg. Examine the foot, finding the V-shaped frog in the center, surrounded by the horny hoof.

Find the pulse by passing the fingers downward from the upper curve of the neck, along the inside of the jaw; count the pulse. Notice the position and motion of the ears with their lining of hair, and the position of the eyes, the form of the pupil, and the probable range of vision. Watch the horse use its lips, and examine the mouth and teeth, finding the grinding teeth far back in the mouth, the incisors in front, and the space where the canines are missing.

The male may have canines in the upper jaw.

On the surfaces of the incisors are the depressions, or "cups," by means of which age is determined.

At six years the cups leave the lower center teeth; at seven the adjoining teeth; and at eight, the outer lower teeth. At nine years they leave the upper center incisors; at ten, the adjoining teeth; and at eleven, the outer teeth above. At the age of ten years a spot appears in the outer upper incisors, at fifteen years the groove has worn to the center of the tooth, and at twenty-one years the groove is worn to the bottom of the tooth.

Questions.

1. Describe the horse you studied as to its name or number, its color, markings, weight, and size. Of what type and breed is it a specimen?

2. Upon how much of the foot does the horse walk? How does this affect ease or speed of action? How does an athlete imitate this in sprinting?

3. How many toes has each foot? What advantage or disadvantage can you see in this unusual structure?

4. How is the hoof constructed to distribute the weight over a surface broader than the leg? How general is this among terrestrial animals?

5. What is the difference in the position of the chestnuts of the fore and hind legs?

6. Where in the foreleg is a springiness permitted by curvature? Where does the back leg accomplish the same thing?

7. How do you account for the elongating of the face?

8. Explain the uses of the lips, telling how they are fitted for their work.

9. Tell where the bit lies in the horse's mouth, and how the structure permits this.

10. Where are the ears situated? How does this peculiar position affect the range of hearing and general alertness? Of how much movement are they capable? Describe the lining of the ear, and state its use.

11. What is the rate of the pulse?

12. Measure the height at shoulder and at croup, length of body from withers to rump, of head, of neck; thickness of body from the shoulder to the chest and of distance of chest from ground. Point out any equalities or ratios you find.

Topics for investigation.

1. The meaning of the terms gee, haw, nigh, off, run, gallop, trot, pace, single foot, rack.

2. The location, cause, and effect of these troubles: heaves, blind staggers, knee sprung, shoe boil, quitter, ring bone, spavin, capped hock, flat foot, hoof bound, glanders, mange, sweeny, hide bound, and thrush.

3. The record time for a trotted and a paced mile.

4. The meaning of "one horse power." How much a horse can pull on good roads.

5. Record prices for valuable horses.

6. Current prices for horses; for ponies; for mules.

7. The origin and the use of the mule.

8. Balanced rations.

9. The number and care of the young, and their relative development at birth.

10. Other animals used as beasts of burden in peculiar conditions or localities.

Homology of the Vertebrate Skeleton

Materials.

Prepared skeletons of an amphibian, a reptile, a bird, another mammal, and man. If any of these be lacking, lantern-slide illustrations may be used in a partially darkened room.

Observations.

Having studied the frog's skeleton in detail, the student can readily compare each of these types with it. Compare in a very general way the skulls, the girdles, and the limbs; their form and use. Note variations in the form and number of the vertebrae and the number of the ribs.

Questions.

1. In which types of vertebrates are the joints between the skull bones bound with cartilage? How does the joining change in later types?

2. What dissimilarities occur in the series as regards closure or boxing in of the eye orbits, nostrils, and skull bones? How would these changes in joining and closure affect strength, rigidity, and protection?

3. What evidence is there that such improvement has affected brain capacity and intelligence?

4. State how the attachment of the skull to the vertebral column changes as the animal man assumes an erect position.

5. Are the vertebrae of these types alike in structure? What is the general form of an horizontally placed vertebra, as in the horse or a reptile, and of a vertically placed one, as in man? If you see any differences, account for them.

6. Wherever possible, find the vertebrae of the neck (cervical), and note the number of them in each case.

7. How is flexibility of the column accomplished in certain types or in certain places of one type? How is rigidity gained? State instances in each answer.

8. Examining the interior of the turtle's "shell," find out and explain how the vertebrae have been modified to form the upper "shell." How has the under portion (plastron) been formed?

9. In round numbers, which skeleton has the greatest number of vertebrae and which the least?

10. Which skeleton has the greatest number of ribs, and which has the least?

11. In a summarizing statement explain any variations you find in the pectoral and pelvic girdle for strength (rigidity); flexibility. This answer may be written as a table, naming the bones, opposite each stating its condition, and then what it affords or is adapted to.

12. What is accomplished by having two bones in the shank of the leg? In what types or forms is there but one, and which one is it?

13. Can you assign any advantages in power, agility, length of leg, or position of leg and foot accruing from a long ankle? (See horse, frog, et al.)

14. Enumerate the types or forms, and opposite each state the number of fingers and toes.

15. Make a table, heading one column "Form or type"; another, "Habitat"; and a third, "Habit." Judging from the structure which you see or from your previous knowledge or experience, fill in the table, stating whether the type is aquatic, terrestrial, or aerial; whether it burrows, walks, runs, or climbs, etc.

16. From your statements in 15, explain how the peculiar mode of life affects the structure of these types.

CHAPTER VII

ADAPTATIONS FOR THE PRESERVATION OF THE SPECIES

#A. METHODS OF REPRODUCTION#

1. The Simple or Asexual Method of Reproduction

Materials.

Slides or diagrams, showing a dividing amoeba, a dividing paramecium, a dividing vorticella, reproduction in some form of sporozoa, budding hydra, gemmules of spongilla, and some species of worm as Dero or Nereis in the process of dividing.

Definitions.

Spore, a cell capable of developing into a new organism.

Asexual reproduction, reproduction by division of the cell or body.

Sexual reproduction, reproduction by means of the conjugation of two reproductive cells known as the egg and sperm cells.

Fertilization, the fusion of the male or sperm cell with the egg or female cell.

Ovary, an organ producing eggs.

Spermary, an organ producing sperm cells.

Cross fertilization, fertilization in which the sperm and egg cells are produced by different individuals.

Dioecious, the different kinds of reproductive organs found in different individuals.

Monoecious, the different kinds of reproductive organs found in the same individual.

Directions.

Note.--Refer to your notes, if the animals mentioned in these exercises have been already studied. This exercise may be largely review.

Study the methods of reproduction in the specimens or diagrams before you. Determine first, in what respects the methods of reproduction are similar in all; second, in what respects there is a variation.

Questions.

1. What has an amoeba gained by dividing? What powers has each new cell that the original amoeba had lost? What would have been the fate of the amoeba if it had not divided into new cells?

2. What various forms of cell division did you find?

3. What is the simplest method of reproduction?

2. The Complex or Sexual Method of Reproduction

Materials.

Slides or diagrams showing hydra and sponge reproducing sexually. Conjugating paramecia, fertilized and unfertilized starfish eggs.

Directions.

Identify the reproductive organs or gonads of the hydra. These are slight swellings on the surface. The one nearer to the mouth end is the spermary and that near the attached end is the ovary.

Questions.

1. How is an egg cell produced in hydra? In general how do the reproductive cells of sponges and hydra originate?

2. What is gained by limiting the process of reproduction to special cells?

3. What is the difference between the appearance of the nucleus of the fertilized and the unfertilized egg?

4. What is the advantage of cross fertilization? How accomplished in Hydra? What reason can you suggest for the spermary's position?

5. Describe the conjugation of a paramecium.

6. Describe the process of maturation and fertilization in a starfish egg.

Suggested drawings.

a. Diving Amoeba or Paramecium.

b. Budding Hydra.

c. Hydra showing gonads.

d. Starfish egg--fertilized and unfertilized.

Summary of important points in the study of methods of reproduction:--

1. What are the two principal methods of reproduction? How do they differ?

2. Why is reproduction necessary?

#B. DEVELOPMENT#

1. The Hen's Egg

Materials.

Hen's eggs, hydrochloric acid.

Definitions.

Germ spot, a white spot, imbedded in the "yolk." This is the point at which development begins.

Yolk, the yellow portion of a bird's egg. This is a food material, rich in fat.

Albumen, the white, viscous portion of a bird's egg.

Chalaza, the spiral portion of albumen always seen in the bird's egg.

Directions.

Boil an egg at least ten minutes in water deep enough to cover it. Note which side is uppermost and mark this part of the shell for reference. Remove the egg and pick away the shell from about half the egg, leaving the shell on that portion which was underneath when placed in water. With a sharp knife remove this half of the egg. Note the thickness of the shell. Test its composition (use hydrochloric acid). Find the membrane lining the shell and note that at one end it separates into two parts to inclose an air space.

1. What is this for? How does it change after an egg has been incubated for a week or more?

Break an uncooked egg in a saucer. Note the germ spot. Note also the difference in the consistency of the "white" (albumen) and yellow portion ("yolk").

2. Why does the latter retain its shape?

3. Why do the white and yolk not mix unless shaken or beaten together?

4. Look for the chalaza.

5. What do you think is the use of this structure?

Weigh a fresh egg, place it in a dry atmosphere for a week, and weigh it again. Record result. Why may eggs be kept a long time perfectly fresh if coated with paraffin or if put in "water glass"?

Suggested drawings.

a. The egg as it appears in the saucer.

b. The egg after part of the shell has been removed.

2. Early Embryonic Development of an Egg

Materials.

Slides or diagrams, showing various stages in the development of some animal through the gastrula stage.

Definitions.

Cleavage stages, two, four, eight, sixteen cells, etc., arising by repeated division, starting with the egg.

Blastula, a hollow sphere, the wall of which is composed of a single layer of cells.

Gastrula, a stage formed from the blastula by pushing in one side of the latter, so as to form a more or less cup-shape structure.

Observations.

1. Is there any considerable difference between the size of the egg and the size of the blastula and gastrula? Has development taken place by an increase of size or by an increase of complexity?

2. Contrast the blastula and gastrula as to number of cavities, number of cell layers, number of external openings.

3. Suggest protozoans that resemble the egg and blastula respectively. What

invertebrates resemble the gastrula in body plan?

Suggested drawings.

a. Some of the cleavage stages.

b. A blastula.

c. A gastrula.

3. Postembryonic Development or Metamorphosis of a Mosquito

Materials.

Some specimens of the larvae and pupae of the mosquito, ordinarily known as wrigglers. Either specimens or diagrams of egg packets should also be provided. Mounted specimens of adult of both sexes.

Definitions.

Postembryonic development, the changes taking place in the development of an animal after birth or hatching.

Larva, the active feeding stage. It is the first stage in postembryonic development, and follows the gastrula stage.

Pupa, usually a resting or quiescent stage. It is the stage following the larva stage.

Observations.

1. Describe the appearance of the egg packet both as seen with the unaided eye and with a hand lens. Find the trapdoor.

2. What is the difference between the appearance of the larva and the pupa? How do their resting positions differ?

3. What does the larva do when disturbed? Describe any characteristic

motions that you notice.

4. Contrast the pupa with the larva under the same conditions and note any differences.

5. Where do you think the external openings of the respiratory organs of the larva and those of the pupa are located? Give reasons for your answer.

6. Into what does the pupa change? Where must the pupa be at this time? Is it easier for the pupa to stay at the surface or at the bottom?

7. Examine an adult mosquito. State the kind of mouth parts, the number and appearance of the wings, the appearance of the antennae. How does the male and female differ in this respect?

Suggested drawings.

a. A diagrammatic drawing representing the jar of water and showing the various positions assumed by the wrigglers.

b. Careful drawings of each stage.

4. Postembryonic Development or Metamorphosis of a Butterfly or Moth

Materials.

The eggs, caterpillars, pupae, cocoons, and adult of some moth or similar stages of a butterfly.

Definitions.

Prolegs, short, unsegmented appendages found in the larva.

Caterpillar, the larva of a moth or butterfly.

Spiracles, openings into the trachae or breathing tubes.

Tubercles, knob-like projections.

Chrysalis, the pupa stage of a butterfly.

Cocoons, the covering spun by the larva before changing to the pupa.

Observations.

1. State the color and appearance of the larva. If tubercles or spines are present, state where. Where are the spiracles? How do you distinguish the head, thorax, and abdomen in the case of the caterpillar?

2. What kind of mouth parts has the caterpillar?

3. How many prolegs has the caterpillar? Of what use are they?

How does the number of prolegs differ from that of the caterpillar in the case of the grub? maggot? currant worm?

4. What is the advantage of the cocoon? What is its color, appearance, and material? Is it composed of a tough substance, or is it easily torn? Where are cocoons found out of doors? Where would you look for chrysalids? (Explain differences in locations.)

Describe the cocoons of tussock moth, clothes moth, leaf roller.

5. What is the difference between a Cecropia cocoon which contains a living pupa and one that has been parasitized? What is the appearance of a parasitized caterpillar?

6. What rudimentary structures can you identify in the pupa? What kind of mouth parts has this stage?

7. Which of these stages is the active stage? Which stage is quiescent? What is really going on in the quiescent stage? In which stage does the insect grow? feed? If the insect were harmful, in which stage would it do the damage? How?

8. Describe what occurs when the pupa changes into the adult. What is the appearance of the wings in the beginning? What changes take place?

9. What kind of mouth parts has the adult? Describe.

10. Describe the antennae.

How do the antennae of moths and butterflies vary?

11. What is the size and appearance of the eggs? Are they laid singly or in groups?

Suggested drawings.

a. A caterpillar, side view.

b. A pupa, ventral view.

c. An adult, dorsal view.

d. A few eggs.

5. Development of the Chick

Materials.

An incubator, a brooder, a setting of eggs.

Directions.

Read carefully the directions for setting up and regulating the incubator. Remember that the temperature should average 103 degrees and should not vary more than two or three degrees above or below this. Candle the eggs from time to time and note difference in appearance, as development proceeds, especially as to transparency and size of the air space. An egg that is transparent after ten days is probably infertile and should be removed. Eggs which are developing properly will show from this time on a well developed air space, and will be quite opaque. The veins often give a spider-web appearance.

Twice each day remove the tray of eggs and allow to cool slightly. Once a day the eggs should be turned before the tray is returned to the incubator.

Questions.

1. Describe the appearance of the shell when the chick is about to come out. In about how many days after you put the eggs in the incubator did you first note this change? You should watch the incubator carefully from the eighteenth day on.

2. What proportion of the eggs hatched? What reasons can you suggest for the failure of some to hatch? Did any which were pipped fail to hatch? If so, break them open and see if you can discover the trouble.

3. How does the chick get out of the shell? How long does it take for it to get out after the shell is chipped? How does a chick look as soon as it has come out of the shell?

4. With what is a chick covered when it is first hatched? How long before feathers begin to develop?

5. How long before a chick needs food? How does it recognize it?

Summary.

1. What are the two kinds of development?

2. What are the stages in embryonic development?

3. What two general types of postembryonic development as determined by the amount of the food supplied in the egg?

#C. PROTECTION AND CARE OF YOUNG#

Library Exercise

Materials.

Books and diagrams showing as many methods for the care of young as possible.

Observations.

1. To what extent is care given to the eggs and young in the case of fish? Is the number of eggs large or small? Contrast this condition with that of the robin. What general conclusion is suggested?

2. Describe the egg-laying habits of five or more of the following: grasshopper, cecropia moth, tussock moth, leaf miner, case bearer, leaf roller, sphinx moth, gall insects, ichneumon flies, spiders, earthworm. How is the developing animal protected in each case? How is food assured?

3. Contrast the method of the honeybee and the solitary wasp as to the method of caring for and feeding the young.

4. Describe the nests of five or more of the following: spider, honeybee, bumblebee, paper wasp, mud dauber, digger wasp.

5. Describe the nests of the following: stickleback fish, sunfish.

6. Describe the nests of ten birds common to your neighborhood.

7. Describe the homes of the following: woodchuck, mole, squirrel, rabbit, muskrat, prairie dog, beaver, bear.

8. In case of birds which of the parent birds builds the nest and cares for the little birds?

9. How are the young cared for in the following cases: crayfish, cyclops, pipefish, Surinam toad?

Summary.

1. What general methods are there for protecting the young?

2. What various devices for assuring plenty of food for the developing animal?

3. What is the relation between the care given the young and the number of eggs produced?

#D. ADAPTATIONS FOR THE PRESERVATION OF THE SPECIES#

Review Questions and Library Exercise

1. Show how the sexual method of reproduction tends to produce variations.

2. What is meant by the term heredity?

3. What are chromosomes? What do some zoologists believe to be the relation between these chromosomes and heredity?

4. What are dominant and recessive characters? What is meant by "Mendel law of heredity"?

5. What is meant by the term parthenogenesis? What are some of its advantages and under what conditions does it take place? Name some animal in which parthenogenesis commonly takes place.

6. What is sex dimorphism? Give some examples.

7. What were the experiments of Professor Loeb and others in connection with artificially fertilized eggs?

8. How do eggs vary as to the kind of shell, amount of food, size, etc.? What is the effect of the amount of food upon the rate of development? On the stage of development at which the egg is hatched?

9. Contrast praecocial and altricial birds.

10. What is the effect of ground nesting and tree nesting upon the number of eggs and the care of the young?

11. Describe the metamorphosis of grasshopper, June beetle, honeybee, dragon-fly, cicada, may-fly, ant-lion, caddis fly.

12. Compare the development of the crayfish, crab, and lobster. What names are given to the larvae? What is the significance in the fact that the lobster hatches in the "mysis stage"?

13. What are some of the peculiar names given to the larvae in the case of echinoderms, worms, and mollusks? Why should these have received special names?

14. Name the three primitive germ layers. State the principal organs derived from each in the higher animals.

15. What is ontogeny? phylogeny? What is the meaning of the law "The ontogeny is an epitome of the phylogeny"?

16. Who was Weissman? What important contribution did he make to zoology?

17. With what phase of zoology is the name of T. H. Morgan associated?

CHAPTER VIII

POULTRY

Materials.

Either pictures or specimens of the different breeds of fowl.

General Information.

Under the term poultry are included chickens as well as turkeys, pigeons, ducks, geese, etc. Chickens are most generally raised, since they do not require such special conditions as the others. In this exercise only this form of poultry is considered.

The hen has been domesticated from prehistoric times, being probably derived from the so-called jungle fowl of India (Gallus bankiva), which is still to be found in its native habitat. Through constant attempts to improve the domestic fowl along the lines of greater egg-production, size, etc., there have been developed a great many different breeds of fowl. These may be divided into seven groups, as follows:

1. The American Class. 2. The Asiatic Class. 3. The Mediterranean Class. 4. The English Class. 5. The Dutch Class. 6. The French Class. 7. The Ornamental and Exhibition Classes.

The American Class includes fowls raised both for egg-production and for eating. It includes the following well-known breeds: the Plymouth Rocks, the Wyandottes, the Rhode Island Reds, and the less-known breeds of Javas, Dominiques, and Jersey Blues. These all lay good-sized brown eggs, are good winter layers, and stand confinement well. The standard weight varies from six and one half pounds to seven and one half pounds for the hen, and from eight and one half pounds to nine and one half pounds for the cock, the Plymouth Rocks being the heaviest of the breeds.

There are three principal varieties of Plymouth Rocks--the Barred Rocks, with grayish-white plumage regularly crossed with bars of blue-black, the White Rocks, and the Buff Rocks. All have single upright combs, which, with the wattles and the ear lobes, are bright red, a large bright eye, and yellow legs.

There are also three principal varieties of Wyandottes. The Silver-laced Wyandotte has a silvery-white plumage, with black markings in various parts of the body. The Golden Wyandotte is similar in its markings, but has yellow where the Silver-laced has white. The White Wyandotte is pure white. All have rose combs, red ear lobes, and yellow legs. They are on the average about a pound lighter than the Plymouth Rocks.

The Rhode Island Reds are a much more recent breed that has of late become very popular. They are of a reddish-brown color, about the weight of the Wyandotte, with yellow legs. There are both single combed and rose combed varieties.

The Asiatic Class includes those breeds raised chiefly for the table. The Brahmas, Cochins, and Langshans are the chief breeds. They are considerably heavier than other breeds, and are specially characterized by the feathers on the legs and feet. They all lay brown eggs, and are in many cases good layers.

The Brahmas include two principal varieties, the light and the dark. The general color is black and white, and they have yellow legs, red wattles, ear lobes, and comb, the latter being of the kind called a pea-comb, which is of small size in the cock.

There are four varieties of Cochins, the Buff being much more raised than the Partridge, the White, or the Black. The Partridge somewhat resembles a dark Brahma, but has red and brown plumage. Cochins have single combs, yellow legs, and a general fluffy character to the plumage, that of saddle and hackle meeting, thus giving a characteristic appearance to these fowls. The eggs are not quite so large as the other two breeds of this class.

The Langshans are smaller and more active than the two breeds just described. They have black legs, the feet are not so heavily feathered, and in general these fowls are much less awkward in appearance. There are two varieties, the White and the Black.

The Mediterranean Class includes those breeds raised chiefly because of their great egg-production. They are active birds, often troublesome because of their ability to fly over high enclosures, so that when kept in the city it is usually necessary to clip one wing. They are not so good winter layers as a rule, but are non-setters. They all lay white eggs. The chief breeds included are the Leghorns, the Minorcas, and the Black Spanish.

The Leghorns--the most popular of these breeds--include two chief varieties, the Brown and the White. The comb is most commonly single, falling to one side in the hen, the wattles long and pendulous, the ear lobes white, and the legs yellow.

The Minorcas are glossy black in color, with a large drooping comb in the hen, and long, thin, pendulous wattle. They lay a very large egg.

The Black Spanish resemble the Minorcas, but are distinguished by the white face and cheeks and the white on the inner edge of the wattles.

The English Class includes the Orpingtons and the Dorkings. The Dorkings are one of the oldest breeds of fowl, and sufficiently identified by the presence of a fifth toe. There are three varieties--the White, the Silver-gray, and the Colored. The White Dorking has a rose comb; the Silver-gray has a single comb and silvery-gray plumage with black markings, the hen having a salmon-colored breast; the Colored Dorkings have sometimes single and sometimes rose combs, the plumage of the cock being black and straw-colored and that of the hen being black and gray with the breast salmon marked with black.

The Orpingtons are short legged, with close plumage. They are of large size, the hens being from seven to eight pounds and the cocks from nine to ten pounds. There are three varieties--the black, the buff, and the white. The black, except for shape, might be mistaken for a Minorca, but has red ear lobes and black shanks. The Orpingtons have the reputation of maturing early, some strains being known to lay when four months old.

The Dutch Class includes the Red-caps, the Campines, and the various varieties of the Hamburgs--of which there are six: the Golden Spangled, the Silver Spangled, the Golden Penciled, the Silver Penciled, the Black and the White. They are all good layers and non-setters, "but lay a small egg, white in color. They are readily recognized by their peculiar rose comb with its long, spikelike projection in the back, their red face, white ear lobes, and bluish legs. The prevailing color of the golden varieties is a reddish bay marked with black and of the silver varieties white marked with black. The cock usually has more dark markings than the hen.

The Red-caps are large fowl with a red and black plumage. The comb is similar to the Hamburg's but larger, and the ear lobes are red.

The Campines resemble the Hamburgs, but have a single comb.

The French Class includes the Houdans, the Crevecoeurs, and the La Fleche. The Houdans are mottled black and white with pinkish white legs, with a fifth toe like the Dorkings, and are easily recognized by their peculiar crest.

The other breeds of fowl, like the crested Polish, Bantams, and game fowl, have less interest for the poultry raiser, though often seen in exhibitions and poultry shows.

To sum up, we may group all these breeds according to their value into (1) the egg breeds, including the Leghorn, Minorca, Spanish, and Red-cap; (2) the meat breeds, including the Brahmas, Cochins, and Langshans; (3) the general purpose breeds, including the Plymouth Rocks, Wyandottes, Rhode Island Reds, Dorkings, and Orpingtons; and (4) the fancy breeds, including the Polish, Bantams, Games, etc.

Definitions.

American Standard of Perfection, an illustrated volume published by the American Poultry Association, indicating the desirable points of each recognized breed of fowl.

Comb, the fleshy outgrowth on the head.

Single comb, a thin, upright comb.

Rose comb, a flat comb with a rough or corrugated surface.

Pea comb, resembling three single combs, united at the back.

Wattles, the fleshy outgrowths from the underside of the throat.

Ear lobes, the fleshy structure in the region of the ear.

Under-color, the color noted when the coverts are raised.

Hackle, the feathers on the neck.

Cape, the feathers back of the hackle.

Saddle, the feathers in the posterior region of the back.

Sickle, the curved feathers of the tail in the cock.

Penciling, small stripes or color markings on the feather.

Spangling, large spots or splotches of color on the feather.

Shanks, the exposed scaly portions of the legs, usually spoken of as the "legs."